T0210740

SpringerBriefs in Molecular Science

Biobased Polymers

Founding Editor

Patrick Navard

Published under the auspices of EPNOE*, *Springerbriefs in Biobased polymers* covers all aspects of polysaccharide sciences, starting from their production and isolation from native sources (i.e. biosynthesis, genetics, agronomy, plant cell biology, biorefinery), over their characterization and processing (chemical / enzymatic modification, shaping, biodegradation) to the many applications in which they are used (food & feed, materials & engineering, biomedical).

The focus of this book series lies on publications related to all kinds of native or synthetic polysaccharides, polysaccharide-derivedpolymers, and composites containing polysaccharides as a fundamental component. Moreover, topics related to natural fibres, wood, polysaccharide containing biomass and bioplastics, life cycle assessments are within the scope.

Under the editorship of Nicolas Le Moigne, Li Shen, Martin Gericke, Stefan Spirk, and Rupert Kargl, the series will include contributions from many of the world's most authoritative polysaccharide scientists, both from academia as well as industry. Readers will gain an understanding on the different strategies developed to obtain these biopolymers and what they can be used for. They will also be able to widen their knowledge and find new opportunities due to the multidisciplinary contributions (plant biology, biotechnologies, bio-based chemistry, wood science and polymers & materials science).

This series is aimed at advanced undergraduates, academic and industrial researchers, and professionals studying or using bio-based polymers. Each brief will bear a general introduction enabling any reader to understand its topic.

* *EPNOE The European Polysaccharide Network of Excellence (www.epnoe.eu) is a research and education network connecting academic, research institutions and companies focusing on polysaccharides and polysaccharide-related research and business.*

Tanja Zidarič · Karin Stana Kleinschek ·
Uroš Maver · Tina Maver

Function-Oriented Bioengineered Skin Equivalents

Continuous Development Towards Complete Skin Replication

 Springer

Tanja Zidarič
Institute of Biomedical Sciences
Faculty of Medicine
University of Maribor
Maribor, Slovenia

Uroš Maver
Institute of Biomedical Sciences
Faculty of Medicine
University of Maribor
Maribor, Slovenia

Department of Pharmacology
Faculty of Medicine
University of Maribor
Maribor, Slovenia

Karin Stana Kleinschek
Institute of Chemistry and Technology
of Biobased Systems
Graz University of Technology
Graz, Austria

Tina Maver
Institute of Biomedical Sciences
Faculty of Medicine
University of Maribor
Maribor, Slovenia

Department of Pharmacology
Faculty of Medicine
University of Maribor
Maribor, Slovenia

ISSN 2191-5407   ISSN 2191-5415 (electronic)
SpringerBriefs in Molecular Science
ISSN 2510-3407   ISSN 2510-3415 (electronic)
Biobased Polymers
ISBN 978-3-031-21297-0   ISBN 978-3-031-21298-7 (eBook)
https://doi.org/10.1007/978-3-031-21298-7

This Springer imprint is published by the registered company Springer Nature Switzerland AG
The registered company address is: Gewerbestrasse 11, 6330 Cham, Switzerland

Preface

Biomedical research relies heavily on cell-based in vitro assays and models to facilitate the understanding of specific biological events and processes under various conditions. Although such systems are simplified to capture a specific domain of physiology, their ability to model specific events in human biology is highly valued as they can bridge the gap between in vivo animal studies and human physiology. The quality of in vitro models in terms of their representation of cell behavior in native tissues is critical to our understanding of cell interactions in tissues and organs. New technological approaches in modern biomedicine have fostered sophisticated and robust in vitro models, ranging from scaffold- or hydrogel-based 3D constructs and organotypic cultures to organs-on-a-chip (i.e., microphysiological) systems.

One of the most studied in vitro models is skin, in which various industries (e.g., pharmaceutical, cosmetic, clinical) have a common interest. Artificial in vitro skin models can mimic the healthy state of the skin and simulate skin diseases. The use of 3D skin equivalents has increased in recent decades as novel preclinical testing systems and as an alternative to animal testing in skin-related industries. Currently, there are several established commercial model systems and a variety of in-house produced, 3D in vitro reconstructed skin models for academic research, all with different characteristics, strengths, and weaknesses. In all cases, the main goal is to generate a layered cellular structure with functional barrier properties similar to human skin. The increasing complexity of 3D in vitro reconstructed skin models raises issues that affect their reliability and predictability. While it has been possible to construct complex models that include fibroblasts, keratinocytes, and other cell types (e.g., melanocytes), the final product exhibits an inherent level of inconsistency and atypical behavior that may be unacceptable for certain purposes and applications. Therefore, 3D in vitro reconstructed human skin model with keratinocytes and fibroblasts provides the basic stable and predictable testing platform.

The book *Function-Oriented Bioengineered Skin Equivalents—Continuous Development Towards Complete Skin Replication* aims to provide potential readers with a comprehensive summary of the available information on various in vitro skin models, from historical background to different modeling approaches and their applications. Particular emphasis is placed on presenting the current technological components available for the development of engineered skin equivalents by summarizing advances in cell cultivation, materials science, and bioengineering. Using examples of the current state of the art, we describe the advantages, limitations, and challenges of developing in vitro skin models for successful use in clinical applications and skin-related research.

Maribor, Slovenia	Tanja Zidarič
Graz, Austria	Karin Stana Kleinschek
Maribor, Slovenia	Uroš Maver
Maribor, Slovenia	Tina Maver

Contents

About the Authors

Tanja Zidarič is a Ph.D. researcher at the Institute of Biomedical Sciences at the Faculty of Medicine, University of Maribor (IBS MFUM), where she has acquired extensive knowledge in the field of tissue engineering, overlapping with materials science, biofabrication, cell biology, chemistry, and microbiology. In addition, her research focuses on the development of electrochemical sensors designed for the detection of clinically relevant biomolecules. She is the project leader of her post-doctoral project in which a part was dedicated to the development of an in vitro skin model. Her bibliography consists of 20 units, including more than 9 scientific papers (most of them in the last 5 years of research and 4 of them related to different sensors, 2 conference contributions, one elaborated, one patent and one patent application.

Karin Stana Kleinschek[1] is a professor at Faculty of Technical Chemistry, Chemical and Process Engineering, Biotechnology, and a head of Institute of Chemistry and Technology of Biobased Systems at Graz University of Technology (TU Graz), where she teaches various courses on polymers, surface properties of polymeric materials, biopolymers, biopolymers in medical application. From 2011 to 2015, she was a vice rector for Research and Development of the University of Maribor. From 2004 to 2016, she was the head of the Institute of Engineering Materials and Design, Faculty of Mechanical Engineering, University of Maribor, and is head of the Laboratory for Characterization and Processing of Polymers (LCPP), which is a part of the institute. She is a member of various scientific organizations (as Member of the scientific committee of International Conferences of Polymer Characterization POLYCHAR, Member of the Electrokinetic Society scientific board, Member of the EPNOE-ACS Conference, etc.). She is the vice president for Research of the European Polysaccharide Network of Excellence (EPNOE)—BIC Association. Since 2013, she is a member of the European Academy for Science and Art and from 2014 associate member of Slovenian Academy of Engineering. In addition, she is a thesis supervisor at University of Maribor, University of Ljubljana, University of Nova Gorica, Slovenia and TU Graz and University of Graz, Austria. Her

---

[1] Member of European Polysaccharide Network of Excellence (EPNOE).

field of expertise is surface modification and characterization of polymeric materials with special attention on polysaccharides and its usability in biomedical applications (3D bioprinting and bioinks development, biopolymer composites and nanoparticles preparations). Her bibliography consists of more than 1088 units, including 201 scientific peer-reviewed papers and book chapters, more than 50 lectures as a visiting speaker, as well as more than 500 contributions from scientific conferences, patents, and innovations. Moreover, she implemented and coordinated more than 41 research projects/activities (2007—present), such as EU projects: (Horizon 2020, 7th FP, Era.NET, COST, EU, Structural funds), programs and projects funded by ARRS, activities relating to the development and implementation of the Slovenia and EU higher education and research policy and more research activities funded directly by Slovenian and EU industries.

Uroš Maver[2] is a head of IBS MFUM and an associated professor in the Department of Pharmacology at the Faculty of Medicine, University of Maribor (MFUM). His research interests include topics such as preclinical in vitro models, surface functionalization, drug delivery systems, tissue engineering and wound healing, and molecular-resolution microscopy, especially atomic force microscopy in life sciences. As part of his research, he also leads several joint projects with the University Clinical Centre Maribor (involving various departments, from plastics, dermatology, orthopedics, and ophthalmology). His bibliography includes more than 372 units (including more than 110 scientific papers, 13 invited lectures, 170 contributions to conference proceedings, more than 15 final reports on national and international projects, 17 elaborates, 1 patent, and 3 patent applications) with a proven track record as a project leader and author of 12 publications recognized as outstanding achievement with the highest scientific merit in the last decade. In 2018, he was awarded the Outstanding Scientific Work Award by the University of Maribor. According to the national evaluation of the Slovenian Research Agency (ARRS), he was among the top 20 researchers in the field of biomedicine in recent years.

Tina Maver[3] The desire to improve patient lives led her into scientific waters. After graduating from the Faculty of Pharmacy University of Ljubljana (UL), she worked with the Trauma Clinic of University Medical Center Ljubljana, where she developed advanced wound dressings for wound care. During her Ph.D. studies in Biomedicine (at the Faculty of Medicine, UL), she developed two prototype wound dressings containing immediate and long-term pain relief drugs. Afterward, she was granted a postdoctoral project and expanded her skin pharmacology research with 3D printing technology. Combining the knowledge of skin pharmacology with biomedical engineering is also her future research direction. Her bibliography includes more than 158 units (including 54 scientific papers, 6 invited lectures, 50 contributions to conference

---

[2] Member of EPNOE (European Polysaccharide Network of Excellence); Member of Institute of Palliative Medicine and Care, Faculty of Medicine, University of Maribor, SI-2000 Maribor, Slovenia.

[3] Member of EPNOE (European Polysaccharide Network of Excellence).

proceedings, 12 final reports on national and international projects, 13 elaborates, 1 patent, and 1 patent applications) with a proven track record as a project leader and author of 6 publications recognized as outstanding achievement with the highest scientific merit in the last decade.

Abbreviations

AD	Atopic dermatitis
ADSC	Adipose-derived stem cell
AGDV	Alanine–glycine–aspartic acid-valine residues (sequences)
ALG	Alginate
BASS	Bioengineered artificial skin substitutes
BEC	Blood endothelial cell
bFGF	Basic FGF
BLA	Biologics License Application
BM	Basement membrane
BM-MSC	Bone marrow-derived mesenchymal stem cell
CAD	Computer-aided design
CAM	Computer-aided manufacturing
CBER	Center for Biologics Evaluation and Research
CC	Cutaneous candidosis or cutaneous candidiasis
CCS	Collagen–chitosan
CEA	Cultured epidermal autograft
CK16	Cyclokeratin 16
CLE	Cornified lipid envelope
CMC	Carboxymethyl cellulose
CS	Chitosan
DDM	Decellularized dermal matrix
dECM	Decellularized extracellular matrix
DHT	Dehydrothermal (treatment)
ECM	Extracellular matrix
ECVAM	European Center for the Validation of Alternative Methods
EDC	1-ethyl-3-(3-dimethylaminopropyl)carbodiimide
EGF	Epidermal growth factor
EPC	Endothelial progenitor cell
EPDS	Expansion particulate dermal substitute
ESC	Embryonic stem cell
FDA	U.S. Food and Drug Administration

FFA	Free fatty acid
FGF	Fibroblast growth factor
GAG	Glycosaminoglycan
GelMA	Gelatin methacrylate
GF	Growth factor
HA	Hyaluronic acid
HaCaT	Immortalized human keratinocytes
hBD-2	Human β-defensins 2
hBEC	Blood vessel endothelial cell
hBOEC	Human blood outgrowth endothelial cell
HIF-1	Hypoxia-inducible factor-1-regulated growth factor
hLEC	Lymphatic endothelial cell
HSE	Human skin equivalent
hSF	Human skin fibroblast
HUCPVC	Human umbilical cord perivascular cell
HuDMEC	Human dermal microvascular endothelial cell
HUVEC	Human umbilical vein endothelial cell
IL	Interleukin
IND	Investigational New Drug
iPSC	Induced pluripotent stem cell
KC	Human keratinocyte
KGF	Keratinocyte growth factor
LaBP	Laser-assisted bioprinting
LBL	Layer-by-layer
LEC	Lymphatic endothelial cell
MADM	Acellular dermal matrix microcarriers
MCTS	Multicellular tumor spheroid
MMP	Matrix metalloproteinase (sequence)
MRSA	Methicillin-resistant *Staphylococcus aureus*
MSC	Mesenchymal stem cell
MUTZ-3	CD34+ human acute myeloid leukemia cell line
MUTZ-LC	MUTZ-3-derived Langerhans cells
NFC	Nanofibrillated cellulose
NHS	N-hydroxysuccinimide
NIKS®	Near-diploid human keratinocyte cell line
NRG-1	Neuregulin-1
P123	Poloxamer
PAM	Poly(acrylamide)
PAMPA	Parallel artificial membrane permeation assay
PCL	Poly(ε-caprolactone)
PDGF	Platelet-derived growth factor
PEG	Poly(ethylene glycol)
PEGDA	Poly(ethylene glycol diacrylate)
PGA	Poly(glycolic acid)
PGS	Poly(glycerol sebacate)

PHEMA	Poly(2-hydroxyethyl methacrylate)
PLA	Poly(lactic acid)
PLACL	Poly(L-lactic acid-co-ε-caprolactone)
PLGA	Poly(lactic co-glycolic acid)
PLLA	Poly(L-lactic acid)
PMA	Premarket Approval
PVPA	Phospholipid vesicle-based barrier assay
R&D	Research and Development
RGD	Arginine–glycine–aspartic acid residues (sequences)
RHE	Reconstructed human epidermis
ROS	Reactive oxygen species
SA	Sodium alginate
SDS	Sodium dodecyl sulfate
SF	Silk fibroin
SKALP	Skin-derived anti-leucoproteinase
SLS	Sodium lauryl sulfate
SoC	Skin-on-chip
SS	Silk serine
SSG	Split skin graft
SVF	Stromal vascular fraction
TE	Tissue engineering
TGF-α	Transforming growth factor-α
TGF-β	Transforming growth factor-β
Th2	T helper cells type 2
TNF-α	Tumor necrosis factor α
UV	Ultraviolet
VEGF	Vascular endothelial growth factor
α-MSH	α-melanocyte-stimulating hormone

Chapter 1
Introduction

The human skin is the largest organ in our body and serves as a barrier and inter-face between the body and the environment. Our skin interacts with various stimuli while maintaining a homeostatic balance. When this balance is disturbed, irritation, inflammation, redness, and itching are typical skin responses to certain stimuli [1–3]. From a clinical point of view, skin diseases are not only a symptom of infectious diseases (e.g. chickenpox, measles), but they also significantly affect the quality of life and mental health of the individual [1, 4]. In addition, skin is the only organ of the body on which all surgeons need to operate on. In their daily practice, surgeons need to possess knowledge about its main anatomical and physiological features, which differ between distinct parts of the body. Furthermore, physicians and surgeons have to be increasingly aware of novel intervention and treatment solutions (e.g., skin grafts and skin flaps, wound dressings for different wound types, etc.). Especially for reconstructive surgery, there is a constant need for improved artificial skin constructs, aiming at fully mimicking the native human skin.

A range of artificial skin constructs was developed in the past for this purpose. Since two-dimensional (2D) cell culture systems cannot adequately replicate the structure and function of complex organs, three-dimensional (3D) tissue models are gaining momentum. The importance of in vitro 3D skin cultures and models has existed for almost a century. Therefore, it is no surprise that some of the most advanced in vitro tissue constructs are related to the skin [5–9]. Various 3D skin equivalents, also known as bioengineered artificial skin substitutes (BASS), ranging from organ culture skin biopsies to vascularized organotypic full-thickness recon-structed human skin analogues, have already been established. Initially, they were used as substitutes for native skin grafts, but nowadays, they are gaining a more prominent role in skin-related research [6, 10–12]. Skin models can be categorized as scaffold-free (i.e., spheroids) or scaffold-based platforms, depending on their format. In the scaffold-based approach, cell proliferation and differentiation are supported by biomimetic matrices. Due to their soft nature, hydrogels are well suited for in vitro models of soft tissues, including skin. They are a special class of scaffolds described as water-swollen polymeric networks. They are typically composed of cross-linked

T. Zidarič et al., *Function-Oriented Bioengineered Skin Equivalents*,
Biobased Polymers, https://doi.org/10.1007/978-3-031-21298-7_1

hydrophilic polymers with the unique ability to absorb and store large amounts of water without dissolving but swelling [13, 14]. In recent decades, the trends in skin engineering were focused on studies and the development of physio mimetic skin models to explore their complex properties, discover new drugs and their mechanism of action. On the other hand, clinical applications were focused mostly on skin regeneration in chronic wounds and burns where normal wound healing is impaired [6, 15].

3D human skin equivalents are considered a promising alternative to animal models, especially in toxicology (i.e., genotoxicity testing) [16, 17]. The skin of animals differs in many respects from the human one. In addition to differences in anatomical structure (e.g., mouse skin is thinner due to fewer keratinocyte layers and contains more hair follicles), there are also significant differences in gene expression (leading to various changes in the skin functional properties). Because of this, there are important differences in the composition of the barrier structure proteins that make up the epidermal differentiation complex. In addition, human skin harbours distinct subsets of immune cells and produces different cytokines than animal skin [18, 19]. This is evidenced by the fact that many drugs and therapies are unable to enter clinical trials and be approved by regulatory agencies despite successful animal testing [14]. Mainly due to the ethical issues raised and the introduction of new regulations (e.g., the 3Rs principle—Replacement, Reduction, and Refinement) resulting from the ban on animal testing, new and efficient skin mimetic equivalents are crucial for pharmaceutical, cosmetic and toxicological purposes [15, 20].

Recent advances in biomedical engineering have enabled the development of bioengineered skin models (some are already commercialized). At the same time, research continues to mimic the biological, architectural, and functional complexity of natural skin [15, 21]. 3D cell culture in a microenvironment that simulates the extracellular matrix (ECM) is expected to improve in vitro-in vivo correlations while reducing the need to perform animal testing accordingly to the 3Rs principle. Tissue engineering is a highly interdisciplinary field in which scaffolds and devices that promote cell growth, organization, and differentiation are used to produce functional 3D tissues in order to restore, replace, or regenerate defective tissues [13, 22, 23]. At the heart of tissue engineering is the principle that successful tissue formation requires the synergistic activity of many cell types, not just the isolated effects of a single population. These cells communicate with each other in a 3D system through living and non-living components. This all-encompassing theme of combining cells, scaffolds, and environmental cues, represents a promising tactic to create artificial tissue for fundamental studies, clinical and industrial applications. These engineered models demonstrate that researchers can study the basic mechanisms of morphogenesis, differentiation, and tumorigenesis by replicating physiological and pathological conditions as closely as possible. However, engineered skin substitutes still face several challenges. These include an accurate recapitulation of tissue physiology, scalability to meet clinical needs, and lowering the financial costs, which hamper their implementation into clinical and pharmacological applications [21, 22].

Nevertheless, skin tissue engineering has a long history of large interest. Considering that large cosmetic and pharmaceutical companies are heavily investing in the modeling and development of living skin substitutes [14], it can be said that the skin tissue engineering has a bright future. Continued progress in interdisciplinary fields like material science, biomedical engineering, physiology, and pharmacology (to name just a few) will likely lead to breakthroughs to overcome current challenges and further boost skin tissue engineering.

This book covers the evolution toward accurate modeling of a variety of human skin equivalents, the development of various methods for creating these models, and their use in various areas of skin-related applications. First, a general introduction to skin physiology and function is given. Then, different 2D and 3D platforms for in vitro skin modeling are presented based on recent literature, starting with a brief historical overview. In the fourth part of the book, we focus on the main constituents to develop tissue engineering human skin models by summarizing the advances and challenges in cell cultivation, materials science, and bioengineering. This part includes a list of available materials, cell sources, and additional biological and chemical cues for more accurate engineering of human skin equivalent. The next part of the book describes the commercially available human skin substitutes and their applications in the clinic and in the pharmaceutical and cosmetic industries. Finally, we summarize the potential applications of in vitro skin models. These applications include skin regeneration/reconstruction, basic biological studies (physiological and pathophysiological processes), and product testing.

References

1. Sutterby E, Thurgood P, Baratchi S, Khoshmanesh K, Pirogova E (2021) Evaluation of in vitro human skin models for studying effects of external stressors and stimuli and developing treatment modalities. View 20210012
2. Yun YE, Jung YJ, Choi YJ, Choi JS, Cho YW (2018) Artificial skin models for animal-free testing. J Pharm Investig 48(2):215–223
3. Eming SA, Krieg T, Davidson JM (2007) Inflammation in wound repair: molecular and cellular mechanisms. J Investig Dermatol 127(3):514–525
4. Karimkhani C, Dellavalle RP, Coffeng LE, Flohr C, Hay RJ, Langan SM, Nsoesie EO, Ferrari AJ, Erskine HE, Silverberg JI (2017) Global skin disease morbidity and mortality: an update from the global burden of disease study 2013. JAMA Dermatol 153(5):406–412
5. Dolbashid AS, Mokhtar MS, Muhamad F, Ibrahim F (2017) Potential applications of human artificial skin and electronic skin (e-skin): a review. Bioinspired, Biomimetic Nanobiomaterials 7(1):53–64
6. Choudhury S, Das A (2020) Advances in generation of three-dimensional skin equivalents: pre-clinical studies to clinical therapies. Cytotherapy
7. Randall MJ, Jüngel A, Rimann M, Wuertz-Kozak K (2018) Advances in the Biofabrication of 3D Skin in vitro: Healthy and Pathological Models. Front Bioeng Biotechnol 6:154
8. Venus M, Waterman J, McNab I (2010) Basic physiology of the skin. Surgery 28(10):469–472
9. Savoji H, Godau B, Hassani MS, Akbari M (2018) Skin tissue substitutes and biomaterial risk assessment and testing. Front Bioeng Biotechnol 6:86
10. Sanabria-de la Torre R, Fernández-González AFV, Quiñones-Vico MI, Montero-Vilchez T, Arias-Santiago S (2020) Bioengineered skin intended as in vitro model for pharmacosmetics, skin disease study and environmental skin impact analysis. Biomedicines 8(11):464

11. Sarkiri M, Fox SC, Fratila-Apachitei LE, Zadpoor AA (2019) Bioengineered skin intended for skin disease modeling. Int J Mol Sci 20(6):1407

12. Vijayavenkataraman S, Lu W, Fuh J (2016) 3D bioprinting of skin: a state-of-the-art review on modelling, materials, and processes. Biofabrication 8(3):032001

13. Fitzgerald KA, Malhotra M, Curtin CM, O'Brien FJ, O'Driscoll CM (2015) Life in 3D is never flat: 3D models to optimise drug delivery. J Control Release 215:39–54

14. Moysidou C-M, Barberio C, Owens RM (2021) Advances in engineering human tissue models. Front Bioeng Biotechnol 1566

15. Moniz T, Costa Lima SA, Reis S (2020) Human skin models: from healthy to disease-mimetic systems; characteristics and applications. Br J Pharmacol 177(19):4314–4329

16. Pfuhler S, Fellows M, van Benthem J, Corvi R, Curren R, Dearfield K, Fowler P, Frötschl R, Elhajouji A, Le Hégarat L (2011) In vitro genotoxicity test approaches with better predictivity: summary of an IWGT workshop. Mutat Res/Genetic Toxicol Environ Mutagen 723(2):101–107

17. Pfuhler S, van Benthem J, Curren R, Doak SH, Dusinska M, Hayashi M, Heflich RH, Kidd D, Kirkland D, Luan Y (2020). Use of in vitro 3D tissue models in genotoxicity testing: strategic fit, validation status and way forward. Report of the working group from the 7th International workshop on genotoxicity testing (IWGT). Mutat Res/Genetic Toxicol Environ Mutagen 850:503135

18. Emmert H, Rademacher F, Gläser R, Harder J (2020) Skin microbiota analysis in human 3D skin models—"Free your mice." Exp Dermatol 29(11):1133–1139

19. Rademacher F, Simanski M, Gläser R, Harder J (2018) Skin microbiota and human 3D skin models. Exp Dermatol 27(5):489–494

20. Danilenko DM, Phillips GDL, Diaz D (2016) In vitro skin models and their predictability in defining normal and disease biology, pharmacology, and toxicity. Toxicol Pathol 44(4):555–563

21. Yu JR, Navarro J, Coburn JC, Mahadik B, Molnar J, Holmes JH IV, Nam AJ, Fisher JP (2019) Current and future perspectives on skin tissue engineering: key features of biomedical research, translational assessment, and clinical application. Adv Healthc Mater 8(5):1801471

22. Hutmacher DW, Horch RE, Loessner D, Rizzi S, Sieh S, Reichert JC, Clements JA, Beier JP, Arkudas A, Bleiziffer O (2009) Translating tissue engineering technology platforms into cancer research. J Cell Mol Med 13(8a):1417–1427

23. Ovsianikov A, Khademhosseini A, Mironov V (2018) The synergy of scaffold-based and scaffold-free tissue engineering strategies. Trends Biotechnol 36(4):348–357

Chapter 2
Skin Physiology and Function

The skin is the largest organ of the body (it accounts for about 15% of total body weight in adults) and, together with its derived structure, forms the integumentary system. As the body's outermost layer, the skin is the first line of defence against external stimuli, making it also the most vulnerable part of the body to injury [1–7]. While intact, we generally do not even notice the skin's overall health and take its look and function for granted. Therefore, it is quite expected that skin diseases have a profound effect on the physical well-being and the quality of personal and social life. Namely, the skin is considered a mirror of general health and is indeed a window into inherited connective tissue disorders. Already in traditional medicine, healers often diagnosed various diseases based on the appearance of the skin and its other changes [6].

Mammalian skin (Fig. 2.1) has a complex, multi-layered structure that includes three hierarchical layers (Table 2.1): epidermis, dermis, and hypodermis (hypodermis adipose tissue). Each layer differs in its architecture, biochemical composition, water content, physical and mechanical properties [4, 6–10].

The skin's thinnest and outermost layer is the epidermis, which plays the most crucial part in the skin's barrier function. This layer is avascular and consists of keratinocytes. Still, it also harbours pigment-producing melanocytes and dendritic cells, including antigen-presenting Langerhans cells, which are involved in the host immune response. Epidermal homeostasis is determined by proliferation rate, differentiation, apoptosis, cellular interaction, adhesion and interaction of keratinocytes with the underlying dermis [12, 13]. Considering its lack of direct blood supply, nutrients within this layer are delivered by diffusion through intercellular fluids once they have passed the selective barrier of the basal membrane separating the epidermis from the vascular dermis. The epidermis can be divided into four sublayers, from innermost to outermost: *stratum basale, stratum spinosum, stratum granulosum*, and *stratum corneum*. The *stratum basale* is where keratinocyte differentiation begins and is generally only one cell thick. It also contains melanocytes, which make up 5–10% of the cell population in this epidermal layer. The basal cells migrate to the surface and form a *stratum spinosum* or prickle cell layer. Langerhans cells can be

T. Zidarič et al., *Function-Oriented Bioengineered Skin Equivalents*,
Biobased Polymers, https://doi.org/10.1007/978-3-031-21298-7_2

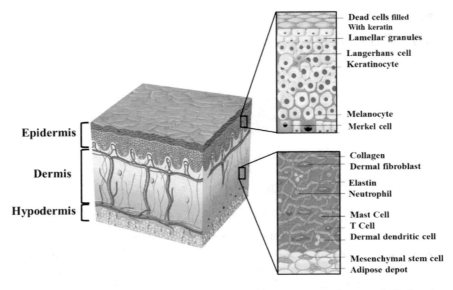

Fig. 2.1 Skin structure showing major components and layers necessary for normal skin function. Reproduced from [11] with permission from MDPI

Table 2.1 Main cell types, ECM components, and functions present in each layer of human skin tissue

Layer	Cell types	ECM components	Functions
Epidermis	Keratinocytes Melanocytes Langerhans cells Merkel cells	Keratin Type IV/VII collagen (basement membrane)	Protection against the external environment and water loss Production of the pigment melanin, protection from UV-B light exposure Sense of touch (associated with Merkel cells) Host immune response (Langerhans cells)
Dermis	Fibroblasts Mast cells	Type I collagen Elastin Proteoglycans Type IV/VII collagen (basement membrane)	Synthesis of ECM proteins Stress-resistance and elasticity Role in wound healing (fibroblasts and mast cells) Allergy response, anaphylaxis, and angiogenesis (mast cells)
Hypodermis	Fibroblasts Adipocytes Macrophages	Type I collagen Elastin	Synthesis of ECM proteins Energy storage in the form of fat tissue Phagocytosis, adaptive immunity, wound healing

Adapted by Refs. [4, 5]

found here as well. In the *stratum granulosum*, keratinocytes secrete lipid components into the intercellular space, contributing significantly to the barrier function and intercellular cohesion within the *stratum corneum*. The latter is the outermost layer of the epidermis and consists of "actually" dead cells (they are without nuclei and cytoplasmic organelles), the so-called corneocytes [1, 3–7, 13]. The *stratum corneum* is a complex, multi-layered formation, traditionally referred to as a "bricks and mortar" structure that provides an effective cellular barrier between the external environment and the internal milieu of living cells. Corneocytes, which are protein-rich cells, represent the bricks embedded in a mortar of a multilayer lipid structure. The wall of corneocytes is composed of highly cross-linked proteins tightly bound to lipids, forming a cornified lipid envelope (CLE). The CLE provides a template for the barrier lipids that bind with the lamellar lipid layers to form the composite structure of the *stratum corneum* [12, 13]. The epidermis is attached to the underlying dermis by a complex network of proteins and glycoproteins that extends from the inner basement membrane into the superficial dermis. In addition, to enhance adhesion, components of the dermal–epidermal junction also contribute to structural support, cell migration, and embryonic differentiation [6, 7].

The dermis is the connective tissue layer beneath the epidermis. It is well connected to the epidermis by the dermal–epidermal junction. It accounts for about 90% of skin weight and forms the foundation of this organ. The dermis is the main source of elasticity, flexibility, and strength of the skin. It is composed mainly of an extracellular matrix (ECM) secreted by its most abundant cells called fibroblasts. Fibroblasts also produce two main protein fibres, namely collagen (75% are type I collagen and 15% are type III collagen) and elastin, as well as glycosaminoglycans (GAG), such as hyaluronic acid (HA). The ECM network of polysaccharides and proteins provides resilience to the skin and a remarkable capacity for retaining water [1, 3, 6, 7, 13]. The dermis is rich in blood supply (although no vessels pass through the dermal–epidermal junction), nerve endings, and various glands [3, 5, 7]. The dermis can be divided into two subcategories, papillary and reticular dermis. The papillary dermis comprises small diameter collagen fibrils and provides the primary protection against mechanical stress. On the other hand, the reticular dermis consists of interwoven bundles of large diameter collagen fibrils that provide elasticity to the skin and have elastic mechanical properties [12–14]. The interwoven collagen and elastin fibrils provide structural support to the skin due to their tensile strength and stretching ability at low forces. The skin has been reported to exhibit highly nonlinear elastic behavior due to the geometric and material nonlinearity of the elastin and collagen fiber network. At low tensile stresses, the skin has a low modulus of elasticity with a linear response mediated by the elastin fibers, while the collagen fibers are still wavy and their stiffness is low. When a certain load is reached, the collagen fibers stiffen and lose their waviness until they become straight, resulting in higher skin stiffness [15, 16].

The hypodermis or subcutaneous tissue is the innermost layer of the skin, composed predominantly of adipose cells that play a key role in energy storage and thermoregulation. In addition, it can be considered part of the endocrine system as it acts on the hypothalamus through the release of hormones such as leptin, which regulates appetite and the control of energy metabolism. The hypodermis is a sophisticated lipid barrier rich in stem cells, hormones, and growth factors (GFs). Because this skin layer is intertwined with nerves and blood vessels that penetrate the upper layers, it contributes to re-epithelialization and wound healing. In addition, adipocytes also have an important signalling function in osteogenesis and angiogenesis. However, the hypodermis is often a neglected compartment in skin models [5, 7, 17].

The skin provides a mechanical barrier against the external environment and performs many other functions. It is involved in innate and adaptive immunity against bacteria, viruses, and fungi through keratinocyte-derived endogenous antimicrobial peptides (defensins and cathelicidins). At the same time, Langerhans cells act as sentinel cells to initiate an immune response against microbial threats and contribute to immune tolerance in the skin. The skin microbiota is another key factor in the protective immune response. The cornified cell envelope and *stratum corneum* reduce the skin's loss of water and electrolytes. The *stratum corneum* also helps against the penetration of irritants and other allergens that can cause skin inflammation. Melanin, found mainly in basal keratinocytes, protects against ultraviolet (UV) radiation. As mentioned, another important function is thermoregulation. It is controlled by vasodilation and vasoconstriction of the skin's superficial and deep vascular plexus. Heat loss is also controlled by sweating. Because of the free nerve endings and terminal corpuscles, it acts as a sensory organ, and its ability to synthesize vitamin D contributes to calcium metabolism and bone formation [3, 7, 17].

References

1. Venus M, Waterman J, McNab I (2010) Basic physiology of the skin. Surgery 28(10):469–472
2. Sanabria-de la Torre R, Fernández-González AFV, Quiñones-Vico MI, Montero-Vilchez T, Arias-Santiago S (2020) Bioengineered skin intended as in vitro model for pharmacosmetics, skin disease study and environmental skin impact analysis. Biomedicines 8(11):464
3. Sarkiri M, Fox SC, Fratila-Apachitei LE, Zadpoor AA (2019) Bioengineered skin intended for skin disease modeling. Int J Mol Sci 20(6):1407
4. Vijayavenkataraman S, Lu W, Fuh J (2016) 3D bioprinting of skin: a state-of-the-art review on modelling, materials, and processes. Biofabrication 8(3):032001
5. Yu JR, Navarro J, Coburn JC, Mahadik B, Molnar J, Holmes JH IV, Nam AJ, Fisher JP (2019) Current and future perspectives on skin tissue engineering: key features of biomedical research, translational assessment, and clinical application. Adv Healthc Mater 8(5):1801471
6. Menon GK, Skin basics; structure and function. In: Lipids and skin health. Springer, pp 9–23
7. McGrath JA, Uitto J (2016) Structure and function of the skin. In: Rook's textbook of dermatology, 9th ed, pp 1–52
8. Haldar S, Sharma A, Gupta S, Chauhan S, Roy P, Lahiri D (2019) Bioengineered smart trilayer skin tissue substitute for efficient deep wound healing. Mater Sci Eng C 105:110140
9. Ramasamy S, Davoodi P, Vijayavenkataraman S, Teoh JH, Thamizhchelvan AM, Robinson KS, Wu B, Fuh JY, DiColandrea T, Zhao H (2021) Optimized construction of a full thickness human skin equivalent using 3D bioprinting and a PCL/collagen dermal scaffold. Bioprinting 21:e00123
10. Pereira RF, Barrias CC, Granja PL, Bartolo PJ (2013) Advanced biofabrication strategies for skin regeneration and repair. Nanomedicine 8(4):603–621

11. Tavakoli S, Klar A (2020) Advanced hydrogels as wound dressings. Biomolecules 10(8):1169
12. Haake A, Scott G, Holbrook K (2001) Structure and function of the skin: overview of the epidermis and dermis. Biol Skin 2001:19–45
13. Suhail S, Sardashti N, Jaiswal D, Rudraiah S, Misra M, Kumbar SG (2019) Engineered skin tissue equivalents for product evaluation and therapeutic applications. Biotechnol J 14(7):1900022
14. Nemoto T, Kubota R, Murasawa Y, Isogai Z (2012) Viscoelastic properties of the human dermis and other connective tissues and its relevance to tissue aging and aging–related disease. In: Viscoelasticity-from theory to biological applications, pp 157–170
15. Daly CH (1982) Biomechanical properties of dermis. J Invest Dermatol 79(1):17–20
16. Guissouma I, Hambli R, Rekik A, Hivet A (2021) A multiscale four-layer finite element model to predict the effects of collagen fibers on skin behavior under tension. Proc Inst Mech Eng H 235(11):1274–1287
17. Lai-Cheong JE, McGrath JAJM (2017) Structure and function of skin, hair and nails. Medicine 45(6):347–351

Chapter 3
Bioengineered Skin Substitutes

The skin can self-renew, thanks to the presence of stem cells in the hypodermis. However, when the skin is damaged in the deeper layers, as in second or third-degree burns, the normal wound healing responses are impeded, resulting in a chronic injury. Consequently, this can lead to considerable disability or even end in morbidity [1, 2]. Skin grafts have long been used as a solution for wound management, but they are associated with limitations in origin source and rejection problems [1–3]. To develop constructs that can mimic and surpass the skin, it is essential to understand its key functions and responses, which are manifestations of its multi-layered structure. As mentioned earlier, skin acts as a protective barrier and as a self-healing material for wound closure that senses changes in pressure and heat. It further transforms in the presence of stimuli, such as light [4]. Bioengineered skin is a natural evolution of the first generation of artificial skins for wound treatment. The ever-evolving techniques of TE enable the improvement and development of new tissue substitutes by incorporating new microstructures and materials to promote cell adhesion and growth to stimulate tissue regeneration. The purpose of bioengineered artificial skin substitutes is to replace or model skin tissue with a construct that mimics its natural physiological form and/or function. Artificial skin is defined as a biological or synthetic substitute for human skin that is produced in vitro in laboratories. Two main compelling reasons have contributed to their establishment and expansion, (i) wound healing and skin regeneration (especially burns and chronic wounds), (ii) drug and cosmetic testing. Despite all the advances and breakthroughs in this field, not all the properties of natural skin could yet be replicated in bioengineered skin analogues, and an ideal skin substitute remains a promise [2, 3, 5, 6].

3.1 A Brief History

The history of skin TE is long and began with the use of skin grafts for wound closure. The first skin grafting procedure can be traced back to 2500 BC in India to

T. Zidarič et al., *Function-Oriented Bioengineered Skin Equivalents*,
Biobased Polymers, https://doi.org/10.1007/978-3-031-21298-7_3

treat severe injuries to the extremities [2, 7, 8]. At that time, it was common to punish a thief or adulterer by amputating the nose, and surgeons of that time took free grafts from the buttocks to repair the deformity [8]. Another millennium passed before the first written record of an original xenograft, using frog skin as a skin substitute, was found in the fifteenth century BC in the Ebers Papyrus [9]. Skin allografts were first mentioned 3000 years later (in the first half of the fifteenth century) in writings of the Branca family of Sicily [9]. After that, nothing significant was done in this regard until the nineteenth century. In 1804, an Italian surgeon named Giuseppe Baronio (in some literature also referred to as Boronio [10]) succeeded in autografting a full-thickness skin graft on a ram [10, 11]. At the beginning of this century, skin grafting went through an important technological development, establishing xenografts. The allografts, a method of taking large films of epidermis with a thin portion of dermis from another person, was demonstrated in 1870, when Sir Astley Cooper grafted a full-thickness piece of skin from an amputated thumb onto a man's stump for coverage. A decade later, the first synthetic grafts were described, and in 1895 Mangoldt introduced the concept of "epithelial cell seeding" [2, 8]. In the early 1900s, human amnion (i.e., a fetal membrane that encloses an embryo) began to be used as a biological dressing to treat burns. Applying amnion to a wound bed prevents desiccation and excessive fluid loss and provides analgesia by protecting exposed nerve ends from the environment. In addition, amnion epithelial cells secret a plethora of regulatory mediators that lead to the promotion of cell proliferation, differentiation, and epithelialization, as well as the inhibition of fibrosis, immune rejection, inflammation and bacterial invasion [12]. Over the course of 40 years, there have been breakthroughs in the ability to bioengineer tissue substitutes, resulting in a wide range of products [9]. Modern attempts to develop human skin substitutes began in the 1960s as a result of advances in tissue culture technology. Karasek presented a method for preparing and maintaining orthotopic grafts from rabbit skin epithelial cells [13]. Cultures of postembryonic rabbit epithelial cells retained the ability to differentiate and form an epidermis after transplantation to an appropriate graft site. One week after transplantation of the cells, the wound site was covered by new growth of epithelial cells and their progeny. After six weeks, the epidermis showed the initial signs of its deterioration. This study paved the way for Rheinwald and Green's success in 1975 when they cultured the epidermis from the patient's own keratinocytes. By cultivating keratinocytes in serial culture, they overcame the limitations of propagating mammalian cells and made it possible to cultivate cells in large quantities [14]. Their work was the revolutionary milestone that has since driven intensive research in the skin TE. Numerous researchers modified the Rheinwald-Green method to produce sheets of autologous keratinocytes [15–17]. Cultured epidermal autografts (CEAs), which emerged in 1981 [15], became the first grafting of the autologous epithelium to cover extensive burns. It was also found that the host tissue used the bilayer to synthesize neoepidermis and neodermis [18, 19]. This laid the foundation for the first commercially available epithelial autografts in 1988 with the development of Epicel®, a cultured epidermal autograft [9]. Variations of the CEA preparation have been used to treat burns, venous ulcers, and other conditions requiring grafting [20,

21]. Because a CEA composed only of keratinocytes is fragile, its application in clinical practice is limited [2]. In parallel with CEA development, a dermal substitute was also introduced to treat extensive burn injuries [2, 18], now known as the Integra® Dermal Regeneration Template [22, 23]. Refinements of the matrix concept [24] contributed to the later development of Dermagraft®, a living, metabolically active, immunogenically inert dermal substitute [20, 25]. In the mid-1980s, the importance of a dermal vascularized component as a bed for CEA formation was recognized [2, 26]. The 1990s was the decade of composite skin substitutes in which both a dermal and an epidermal layer were fused into one substitute [2, 6]. These events led to establishing a new research field, tissue engineering (TE), in 1993 by Langer and Vacanti [27]. In its essence, TE combined engineering design principles and knowledge of biological mechanisms to replace or regenerate damaged tissue [2, 6, 27]. Since then, a wide range of skin substitutes have been produced, usually in the form of allogeneic skin cell populations seeded layer-by-layer on scaffolds of ECM proteins [2, 6]. In the new millennium, this technology has been pushed for commercialization by the "big names" in the cosmetics industry to develop suitable alternatives to animal testing. Just to illustrate, the subsidiary of the well-known cosmetics company L'Oréal owns the Episkin™ product line. Episkin™ and other similar products are reconstructed skin models widely available and widely used as in vitro substitutes for human skin [6, 28–31]. The innovation and "Research and Development" (R&D) departments of major companies have made significant progress in introducing skin models [6].

3.2 2D Cell Culture Models

For over a century, two-dimensional (2D) cell cultures have been used as in vitro models to study cellular responses to biophysical and biochemical stimuli. Although these approaches are well accepted and have contributed remarkably to our understanding of cell behaviour, under certain circumstances, the 2D systems can lead to cellular activities that differ significantly from the response in vivo. The 2D platforms have gained acceptance in skin TE as they are the most common and longstanding representatives of skin modelling. The epidermis is often represented as a monolayer of keratinocytes to study skin permeation and screening of substances [32, 33]. Conventional 2D cell culture relies on attachment to a flat surface, usually a glass or polystyrene Petri dish, to provide mechanical support. Cell growth in monolayers allows access to a similar amount of nutrients and GFs in the medium, resulting in homogeneous growth and proliferation [32–34]. Furthermore, access to nutrients is not affected by a cell gradient, as developed in human tissues, because necrotic cells detach into the medium, leaving only viable cells on the culture surface [32, 34]. However, the environment in which the cells reside is unphysiologically rigid and provides no control over cell shape, determining the biophysical factors that influence cellular activities in vivo. More specifically, this 2D microenvironment impacts proliferation and morphology restricts directional migration and alters biochemical signalling with gene and protein expression. This leads to discrepancies between in vitro and in vivo results [32, 33, 35]. To control cell shape in 2D cell models, micropatterned substrates (cell-adhesive islands, microwells, and micropillars) have been developed to fit the 2D shape of cells. However, these pseudo-3D models induce apical-basal polarity, which is unnatural in vivo for some cell types, such as mesenchymal stem cells (MSCs). This induced polarity can interfere with the functions of native cells in terms of spreading, migration, and recognition of soluble factors and other stimuli from the microenvironment [32, 36]. Although the prevalence of 3D culture systems is increasing due to the ability to replicate the in vivo environment with acceptable accuracy, many epithelial systems can be reasonably approximated in 2D. For example, in an air–liquid interface culture, keratinocytes are stratified on 2D surfaces. Due to the simplicity of the systems, 2D monolayers sometimes do not show the cell development process seen in the physiological environment [32]. To mimic ECM in vivo, which cells see in 2D, collagen or fibronectin is usually used to allow cells to bind to substrates [32, 37, 38]. However, with conventional 2D culture methods, mimicking 3D ECM can be challenging because 2D cell models are limited to a single cell line. Moreover, only ventral adhesion is allowed in 2D in vitro cultures, creating an unnatural environment that differs from natural 3D ECM. This in turn triggers various cell adhesion signaling pathways, thereby modulating important processes such as cell proliferation, differentiation, and gene expression [39, 40]. To compensate for the lack of ECM, several researchers have used a sandwich culture method developed from two 2D (e.g., ventral and dorsal) substrates (Fig. 3.1), in which an ECM layer is applied to the cells. Simultaneous activation of ventral and doral receptors (by overlaying them with a film of a new material) can elicit cell responses that resemble a 3D environment, causing changes in cell morphology, adhesion and intracellular signalling pathways, and cell migration. An advantage of this system is that glass coverslips or various porous (plastic) materials can be used as ventral and dorsal substrates, which can be coated with a thin layer of ECM proteins (usually collagen I and fibronectin). In this way, the morphology and function of the cells can more accurately reproduce behaviour in vivo due to the distribution of cell receptors anchored to the ECM [32, 39–43].

Still, monolayer cells often exhibit different characteristics than 3D cell cultures, including ECM deposition, GF secretion, and gene expression profiles. Thus, they do not map cell–cell or cell–matrix interactions or signalling pathways [32, 33].

3.3   3D Cell Culture Techniques

An important catalyst for the development of skin substitutes was the ever-present scarcity of healthy skin surfaces in patients to cover their wounds in cases of extensive/extreme trauma. Although the painful procedure the patients need to endure in autografts is not helping either. At their core, skin substitutes are a heterogeneous group of materials that contribute to the temporary or permanent closure of many wound types. Their basic characteristics are ideally aligned with respective wound requirements [3, 5, 44]. Most of the skin substitutes currently available are merely

2D   **Sandwich-like culture**

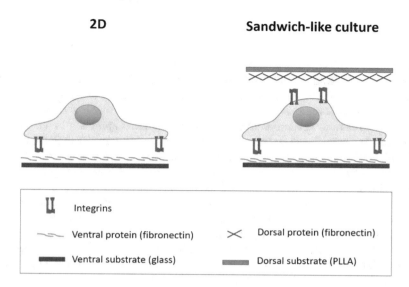

𝐼𝐼	Integrins		
〜	Ventral protein (fibronectin)	✕	Dorsal protein (fibronectin)
▬	Ventral substrate (glass)	▬	Dorsal substrate (PLLA)

Fig. 3.1 Sketch of traditional and sandwich 2D culture methods. Reproduced from [39] with permission from My JoVE Corporation

equivalents of the epidermis and/or dermis composed of primary cells (keratinocytes, fibroblasts, and/or stem cells) and components of the ECM. As aforementioned, epidermis analogues are quite fragile, and keratinocytes exhibit a lower proliferation rate without fibroblasts. On the other hand, dermal substitutes lack the outer protective skin barrier. Full-thickness dermo-epidermal equivalents are more representative and have obviously improved and accelerated healing quality in clinical applications compared to simple dermal or epidermal counterparts. However, they are still a simplified representation of human skin. In contrast, the full structure and function of the skin as an organ depends on all layers, cells and appendages to function properly [2, 45–47].

The use of human cells and ECM components provide the required 3D environment for normal cell growth, proliferation, and function that is more representative than 2D cultures or animal models, without raising ethical concerns [44, 47]. Nevertheless, in vitro pathophysiological studies are still commonly performed using 2D cell cultures. These studies are valuable for gaining greater insight into molecular signalling, cell morphology, and drug discovery, but not all the knowledge gained is transferable to in vivo physiological systems. The cell-to-cell, cell-to-matrix, and cell-to-environment interfaces associated with a 3D environment are critical for physiologically relevant cellular functions. Their absence in 2D models affects cellular responses, from morphology, proliferation, migration and differentiation to biochemical signalling, gene and protein expression [46–49]. Furthermore, primary skin cells cultured in 3D platforms mimic in vivo physiological conditions and allow personalized mechanistic and translational studies [46].

3.3.1 Scaffold-Free 3D Models

The production of skin substitutes by the classical TE approach has relied on creating 3D polymeric scaffolds as ECM analogues to drive cell adhesion, growth, and differentiation to form functional and structural skin tissue. However, using these scaffolds is associated with some risks due to immunogenic reactions and low tissue viability after transplantation. A more enthralling and less studied approach to TE induces fibroblasts to synthesise their own ECM by self-assembling rather than using scaffolds. The resulting skin substitutes are xeno-free, and the risks associated with the use of scaffolds are avoided. A scaffold-free method is a bottom-up approach using either a single cell suspension, spheroid cell aggregates, tissue strands, or cell sheets as building blocks. The process depends on the inherent ability of the building blocks to combine and form larger cohesive structures that produce ECM. Due to the high initial cell seeding density, cell proliferation and migration are not critical factors in the scaffold-free method, which reduces the time required for tissue construction. A notable advantage of this strategy is the ability to create intricate tissue and organ architectures by producing heterogeneous building blocks from different cell types. However, the absence of a physical barrier can impact mechano-transduction, as it can begin immediately after assembly or grafting. Furthermore, the lack of a complete understanding of all the factors involved in these techniques limits their wider use. Two processes characterize the concept of a scaffold-free technique:

(1) self-assembly of cells without the use of external forces, i.e. scaffold-free bioprinting and cell sheet,
(2) self-assembly in response to natural features without external forces, i.e. non-adherent substrates are used to allow cells to perform all processes with minimal interference, such as spheroid formation [50–52].

3.3.1.1 Spheroids

Spheroids, scaffold-free 3D models, are defined as non-adherent cell aggregates formed by the self-assembly of one or more cell types in an environment that prevents attachment to a flat surface. They are also called "microtissues" and can represent the actual microenvironment of cells more accurately than 2D cell cultures. They are self-assembled due to the gravity of monodisperse cells and the presence of membrane proteins (integrins) and EMC proteins, which allow a closer connection between cells. They can mimic many features of organ and tumour tissue characteristics, including the deposition of ECM proteins, cell–cell interactions, and the formation of nutrient, waste, and gas gradients [45, 46, 53]. Typically, they are generated by the hanging drop method, in which droplets of cell suspensions are cultured at a high cell density in inverted cell culture plates under physiological conditions until they form 3D spheroids (Fig. 3.2). In this method, the cells are in direct contact with each other and with the ECM components. Furthermore, it can be modified by adding biological agents or by co-culturing two or more cell types [45, 54].

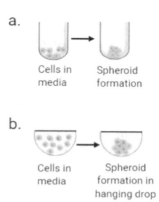

Fig. 3.2 Spheroid culture of skin equivalents: **a** spontaneous formation of spheroids on low adhesion, round-bottom surface; **b** formation of spheroids by hanging drop method. Reproduced from [45] with permission from Elsevier B.V

In skin-related research, 3D spheroid culture is mainly used for cell function, cytotoxicity, or high-throughput biochemical analysis. To achieve satisfactory tissue architecture, cell positioning is critical for producing full-thickness stratified epithelium. To date, a limited number of studies have reported the cultivation of keratinocytes and fibroblasts as spheroids [45, 46]. In their work, Ströbel et al. [55] created a spherical skin microtissue model based on primary dermal fibroblasts surrounded by keratinocytes. The dermal core alone produced the ECM without the need for exogenous collagen for skin formation. The epidermal layer consisted of proliferated keratinocytes in the basal membrane and differentiated keratinocytes in the suprabasal layer. The keratinocytes at the periphery were terminally differentiated and formed the keratinized layer showing a barrier function for penetration [55]. Still, this miniaturized model was completely immersed in the medium, which does not replicate the air–liquid interface of physiological skin and may therefore affect drugs penetration and action [46]. Recently, Klicks and colleagues [56] produced spheroids by co-culturing three cell types (fibroblasts, keratinocytes, and melanoma cells) on a low-attachment surface, in which type IV collagen-rich fibroblasts formed a core, surrounded externally by a ring of keratinocytes and melanoma cells [56]. Even though scaffold-free approaches allow more efficient fabrication of miniaturized skin models and increase the reproducibility of test results, the main application is in anticancer drug development and toxicological evaluation [46, 55].

3.3.1.2 Organotypic Co-culture

An ideal skin equivalent should act as a physical barrier at the interface with the gaseous environment. In the case of the native skin, this barrier is highly dependent on the humidity and culture conditions under which it is created [46]. Scaffold-free cell sheet-based constructs (Fig. 3.3) are successful for tissue repair due to their high nutrient diffusion rates, abundant deposition of ECM, and interactions with cell membrane proteins [45, 57]. In a scaffold-free technique, skin equivalents can be formed by co-culturing fibroblasts and keratinocytes in a layer-by-layer fashion. Keratinocytes are subsequently seeded on cultured dermal substitutes. Most of the co-cultures are then raised to an air–liquid interface to facilitate the stratification of the epithelium [45, 58].

This approach was successfully used to develop the "hairy" human TE skin analogue. The skin substitute was generated exclusively from cultured human fibroblasts and keratinocytes with integrated pilosebaceous units (hair follicles) and without the use of synthetic material. Fibroblasts treated with ascorbic acid formed a well-organized dermal layer due to the increased deposition of ECM proteins (type I collagen and fibronectin). Keratinocytes seeded on the dermal layer fully differentiated and formed a stratified and keratinized epidermis once they were raised to an air–liquid interface. The as-prepared skin model allowed the pilosebaceous units to be embedded in a fibroblast-rich ECM mimicking the microenvironment of the human dermis. The hair follicle remained organized in culture for 5 weeks and expressed various proteins, including hair follicle-specific trichohyalin [58]. Liu and colleagues [57] used a scaffold-free technique and cultured ascorbic acid-treated fibroblasts to develop a vascularized bilayer skin substitute. They co-cultured fibroblasts and endothelial cells to form a dermal sheet with a capillary network. The dermal-specific matrix components secreted by the fibroblasts induced sprouting of the endothelial cells and were necessary for the lumen formation. The capillary-like networks contained typical microvasculature. Keratinocytes were then placed over the dermal analogue to create a bilayer skin equivalent with a capillary network [57].

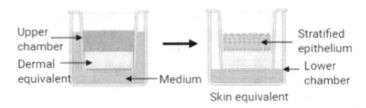

Fig. 3.3 Scaffold-free approach for engineering skin substitute: organotypic co-culture of fibroblasts and keratinocytes on a transwell plate. The upper chamber contains keratinocytes seeded on the dermal equivalent composed of fibroblasts, subsequently raised to an air–liquid interface for the stratification. The lower chamber contains the culture medium. Reproduced from [45] with permission from Elsevier B.V

3.3.2   Conventional Scaffold-Based 3D Models

Scaffold-based 3D models are, in the broadest sense, cells grown in the presence of a support scaffold based on either hydrogels or polymer fibres. Unlike scaffold-free models, these 3D constructs have a better resemblance to the native tissues in terms of structure, mechanical properties and functionality [46]. However, this type of model can affect the long-term behaviour of engineered tissue constructs and directly influence their primary biological functions. This is due to several factors, including immunogenicity, the toxicity of degradation products, scaffold degradation that can cause fibrous tissue formation, and mechanical inconsistencies with the surrounding tissue [57]. Nevertheless, scaffold-based 3D models are generally used, whether for clinical applications or for early evaluation of therapeutic agents or cosmetic products [45, 46, 59, 60]. Considering the biomimetic approach, various scaffold fabrication strategies have been applied in recent decades to match the layered and fibrous skin architecture. The scaffold architecture should mimic the porous and fibrous structure of dermal components and support the formation of the epithelial layer with its primary features. These include keratinocyte stratification, cell polarization, contact with the basement membrane, and effective barrier properties [60, 61]. Building a 3D construct is not sufficient for tissue regeneration/restoration, and the current goal of tissue engineers is to establish a cell-instructive microenvironment [60]. Natural polymers such as collagen, chitosan (CS), hyaluronic acid (HA), gelatin and fibrin are the preferred scaffold materials for developing 3D skin models in vitro [45, 46, 59, 60]. Scaffolds based on ECM proteins are commonly used in vitro to model aspects of skin physiology and transport phenomena and exploit the characteristic properties of protein-based materials. While these models help to capture specific features of native skin tissue, they are generally not comprehensive as they neglect the complexity of skin physiology as a whole [6]. Collagen is a major component of ECM; therefore, collagen-based biomaterials have been mostly used as ECM analogues. The widespread use of different types of collagen in many clinical applications is due to its bioinductive properties, which can support physiological processes in wound healing. Besides, the ability to obtain collagen from natural sources makes it an attractive and physiologically relevant matrix candidate. As a platform for developing functionally intact 3D tissue constructs, it has been reported to promote vascularization both in vitro and in vivo during the reconstruction of skin tissue [45, 46, 62, 63]. However, collagen and other naturally derived biopolymers usually have weak mechanical strength due to their very high-water content (more than 90% [64]). However, since the crosslink density determines the mechanical strength, the stronger hydrogels have a lower ability to absorb water [65]. As for collagen, collagen type I is prone to physical contraction when integrated fibroblasts exert forces on the matrix, usually resulting in relatively short-term in vitro studies. Furthermore, this contracting matrix limits the physiological relevance of the constructs, as more robust systems require matrices that can withstand prolonged experimentation, mechanical handling, and sequential operations to join different layers into a full-thickness system. This problematic contraction can be overcome by removing water from the scaffold through physical compression [46, 66, 67]. To prevent the severe contraction and instability of the hydrogel, Brown et al. [68] developed a plastic compression method that was further optimized by Brazilius and his group [69]. The focus of this process is the rapid removal of water content from the hyperhydrated collagen gel. Reconstituted collagen gels consist of a random meshwork of collagen fibrils with a large amount of excess fluid (99%) due to the casting process (rather than the swelling ability of the collagen) [68]. The initially thick dermal hydrogels are compressed to achieve a final thickness of up to 1 mm before the keratinocytes are seeded. Such collagen sheets typically lose 91% water content. The compressed platform can vary in size, allowing for a more "flexible" fabrication of skin substitutes to meet specific needs [47, 68, 69]. Recently, a new strategy to overcome fibroblast-mediated contractions of collagen-based skin models was proposed by applying keratinocytes to the insert membrane separating the epidermis from the underlying dermis [70]. Another obstacle for long-term cell culture is the tendency for enzymatic degradation, and some of the degradation products can trigger chemotaxis of human fibroblasts. Despite the limitations mentioned above, bioengineered collagen-based skin tissues play a leading role in mimicking healthy and diseased skin in vitro. For this reason, the poor mechanical and thermal properties of collagen shift the field to the use of blends with other (more robust) materials or by implementing additional techniques such as crosslinking [46, 63]. On this matter, a hydrogel composed of cross-linked silk and collagen has been described as an ideal dermal biomaterial [67]. This composition allowed to preserve the cell-binding domains of collagen while benefiting from the stabilizing properties of silk, which is more resistant to time-dependent degradation and contraction than collagen hydrogels [46, 67].

Instead of chemical crosslinking, porous and soft substrates can be generated from naturally derived biomaterials by self-assembly [47, 71], lyophilization or freeze-drying [72, 73], electrospinning [73–75], and knitting [74]. In the self-assembly model, fibroblasts can secrete their own ECM to form a fully autologous dermal equivalent. Usually, these fibroblast-derived ECM sheets are very thin, and the design of the final dermal matrix is done by manually overlaying more than one sheet [47]. Freeze-drying dehydrates hydrogel scaffolds and converts them into dry porous structures, while electrospinning creates porous constructs by attracting hydrogen fibres in between [47, 73]. To facilitate the ingrowth of native cells or the proliferation of seeded cells from autologous or allogeneic sources, these constructs may also contain GFs and cells of interest (usually fibroblasts, keratinocytes or stem cells grown in vitro).

In addition, to increase their efficiency and circumvent the high batch-to-batch variability associated with natural polymers, they are combined with synthetic polymers, including polyesters (poly(ε-caprolactone), PCL, poly(lactic acid), PLA, poly(glycolic acid), PGA, poly(lactic co-glycolic acid), PLGA) and polyethers (poly(ethylene glycol), PEG and PEG co-polymers) [45, 46]. Synthetic polymers are known to be able to adjust their physical properties, such as porosity, biodegradability, and stiffness/elasticity, depending on the desired application. For example, PCL has been used to fabricate electrospun tissue scaffolds, where high porosity and interconnectivity have promoted cell invasion and synthesis of matrix proteins (type I collagen, fibronectin) [76, 77]. However, synthetic polymers often have poor cell adhesive properties. Therefore, in combinations of natural and synthetic polymers, the synthetic polymers mostly determine the robustness of a construct, whereas natural polymers are crucial for the physiological significance of 3D constructs [45, 46]. For example, the architecture and composition of a bilayer nanofibrous PCL and CS (PCL-CS) scaffold promoted efficient cellular activity. The different pore size distribution in the different layers has facilitated compartmentalization and prevented initial cell migration. Moreover, the PCL-CS scaffold has supported ECM protein synthesis and keratinocyte stratification in vitro [78].

3.3.2.1 Reconstructed Human Epidermis (RHE)

Early in vitro skin models consisted of 2D cell culture to test cell viability and immediate cell effects. As development progressed, efforts were directed towards the reconstructed human epidermis model (RHE). In its basic form, RHE consists of a polycarbonate membrane on which keratinocytes are seeded and then cultured at an air–liquid interface to differentiate and form the epidermal sublayers. This 3D reconstruction of the normal human epidermis includes all viable (*stratum basale, stratum spinosum*, and *stratum granulosum*) and non-viable (*stratum corneum*) layers and simulates the skin's barrier function. With this configuration, the researchers could mimic metabolic and biomolecular features that resemble the physiological microenvironment [59, 79, 80]. These RHE models, however, lack dermal components (collagen, fibroblasts, and blood vessels). Like monolayer models, they cannot be used to study the interactions between keratinocytes and other skin-related cells [47, 59]. Besides, other missing epidermal components, such as melanocytes and Langerhans cells, dermal and epidermal leukocytes, limit their immunological repertoire to those derived from keratinocytes alone. Nevertheless, such models have been adapted because of their relative ease of reproducibility and high throughput capacity. Several RHE models have been used extensively to predict topical skin irritation/corrosion and for screening agents in high-throughput settings, which may reduce the need for animal testing [59, 79, 81]. RHE models with melanocytes are used for skin lightening and pigmentation applications [79]. The ban on animal testing of finished products or cosmetic ingredients by the 7th Amendment to the Cosmetics Directive in the early 2000s forced the big names in the cosmetics industry to find alternatives or develop new methods. Consequently, leading cosmetic companies have funded efforts to develop living skin equivalents [6], and the most recognized and widely used RHE models are EpiSkin™ and SkinEthics™ [6, 59, 79]. Both are metabolically and mitotically active, have lipid and ceramide profiles very similar to the normal human epidermis, and express normal epidermal differentiation markers (keratin 1/10, profilaggrin, involucrin). They are cultured in a serum-free medium, and standardized manufacturing processes allow highly reproducible production from batch to batch [59, 79].

3.3.2.2 Human Skin Equivalent (HSE)

In parallel with the development of RHE models, attempts have been made to add a living dermal compartment to create human skin equivalents (HSEs), referred to as "reconstructed skin" or "full-thickness" models [82]. Based on the pioneering work of Bell et al. [83], who developed a dermal equivalent from fibroblasts embedded in a collagen matrix, it was possible to seed keratinocytes directly onto the surface of the formed dermal lattice layer [82, 84–88]. The epidermis is separated from the dermis by the dermal–epidermal junction. Both keratinocytes and fibroblasts are involved in the formation of the dermal–epidermal junction to regulate the process. In the reconstructed HSE model, the dermal–epidermal junction usually has a flat shape, whereas the native skin tissue has an wave-like structure. This undulating structure comes from the epidermal processes migrating downward into the dermis to conform to the shape of the dermal extensions (dermal papilla) [82]. Some HSEs reconstructed on dead, "de-epidermized" dermis exhibited this wavy structure [89]. However, only the epidermis contained the living cells in this case, and such HSEs cannot be produced on an industrial scale due to limited access to the human dermis [82]. The currently existing HSEs are usually fabricated on primary human skin cells, which are very similar to natural skin [87]. In general, an HSE is constructed from normal human skin in the following sequential steps (Fig. 3.4):

(1) First, a thin, acellular collagen layer is built up as an attachment substrate for the cellular collagen matrix produced on top of it. The acellular layer prevents the cellular collagen from contracting and detaching from the insert membrane.
(2) A collagen matrix containing human dermal fibroblasts (a cellular collagen matrix) is prepared and allowed to shrink in the medium for one week. The fibroblasts embedded in the collagen gel remodel the matrix by secreting ECM proteins and causing the gel to contract.
(3) Once contraction and stabilization of the cellular matrix are complete, keratinocytes are seeded onto the surface of the matrix and can attach, forming a confluent cellular monolayer that will induce tissue stratification.
(4) Tissues are raised to an air-liquid interface to enable complete stratification and complete morphological and biochemical differentiation (tissue stratification, organization and differentiation) [86].

Although more complex to fabricate than RHE models, the presence of living dermis broadens the scope of their possible use. The crosstalk between the epidermis and dermis is key to skin homeostasis. Therefore, a model reflecting both epidermis and dermis functions is undoubtedly a driving force in promoting innovative applications [82]. For example, two commercially available full-thickness models, PhenionFT™ and EpidermFT™, have been used for a broader range of applications: environmental and age-related effects [90, 91], skin penetration [92, 93], skin metabolism [94], genotoxicity [95], wound healing [96–98], disease mechanisms [99], and skin sensitization [100].

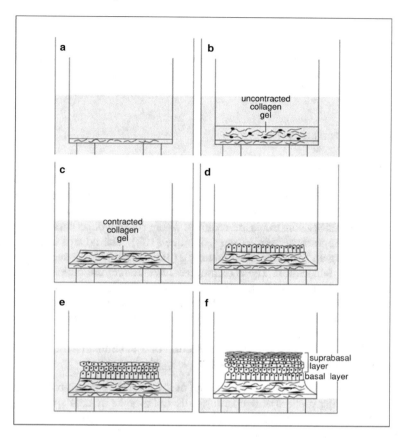

Fig. 3.4 Step-wise construction of HSE model. Reproduced from [86] with permission from John Wiley & Sons, Inc

3.3.3   Biofabrication Approaches

Scaffolds are the backbone of any TE skin equivalent. In recent years, biofabrication technology has arisen as a revolutionary strategy that could greatly improve the outcomes of regenerative medicine and TE. Biofabrication uses manufacturing processes to arrange biomaterials with a hierarchical structure and spatial distribution of one or more cell types and molecules to create complex aggregates that mimic the structure and function of native tissues or organs. Among the various methods, 3D bioprinting and electrospinning are the cornerstones of biofabrication [101, 102]. Both strategies represent powerful tools that enable exquisite control over the microstructure and cytostructure of the bioengineered skin tissue [101, 103]. Moreover, 3D bioprinting facilitates the fabrication of custom tissues with patient-specific geometry through the combined use of medical imaging modalities, computer-aided design (CAD), and computer-aided manufacturing (CAM) [103]. On the other hand, electrospinning is considered as a straightforward and multifunctional technique for the fabrication of ultrafine fibres with diameters in the nanoscale and microscale, allowing a more accurately mimicking of the ECM conditions of the native tissue [104–107]

3.3.3.1 3D Bioprinting

Commercially available in vitro skin models are mainly generated by manual deposition of cells in an appropriate ECM. The major drawback of manual deposition of cells is the inability to precisely position the cells in the ECM. Besides, with the increasing use of 3D skin equivalents in cosmetic surgery and toxicity testing, a larger number of smaller 3D constructs are required where manual deposition becomes an inefficient and impractical fabrication technique [108]. The advent of rapid functional prototyping and additive manufacturing (3D printing) has opened a new field for tissue model fabrication. These techniques have undergone rapid advancement and are used to design complex structures, especially in TE. The computer-controlled positioning of nozzles allows living cells, biological materials, and biochemicals to be precisely deposited layer-by-layer to produce a realistic and robust 3D representation of objects. For skin TE, 3D bioprinting is used to replicate the microfeatures of human skin [3, 45–47, 108]. The construction of a 3D human skin model by 3D bioprinting is commonly executed in a three-step process:

(1) selection of materials, cells, and design of the scaffold before processing,
(2) 3D bioprinting process of depositing cells and bioinks in a 3D structure, and
(3) post-processing, maturation, and characterization of the skin construct [3, 88, 108, 109].

The various 3D bioprinting methods can be categorized according to the two main strategies for biofabrication: bottom-up and top-down (Fig. 3.5). In the bottom-up approach, a larger complex tissue is fabricated by spatially arranging small cellular and acellular building blocks to form a biomimetic complex architecture. The top-down approach typically uses a larger biodegradable scaffold on which cells are seeded and specific GFs are introduced to support cellular activity. Eventually, the cells can synthesize their own 3D matrix to replace the original scaffold [3, 47, 103, 110].

Each strategy has its advantages and disadvantages, consequently reflected in the respective 3D bioprinting methods. The bottom-up approach allows the deposition of multiple cell types with 3D organization. It can enhance vascularization in 3D constructs by generating intra-organic vascular trees to perfuse the constructs and ensure their viability [3]. The top-down approach allows for easy cell migration on the scaffold structure and thus forming cell–cell junctions and better cell–cell interactions. In addition, a wide range of biodegradable and biocompatible materials for scaffold assembly and precise control of pore size and shape, porosity, and pore interconnectivity contribute to structural stability [3, 110]. The main drawback is the

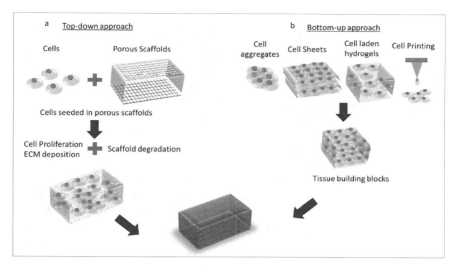

Fig. 3.5 Top-down and bottom-up approach in TE. **a** Top-down approach involves culturing of cells in a porous scaffold. Cells grow, proliferate, migrate on the scaffold structure, and create ECM, while the scaffolds undergo biodegradation, and finally, a matured tissue is obtained. **b** Bottom-up approach uses cell aggregates, cell sheets or cell-laden hydrogels to produce modular or tissue building blocks, which are then assembled together to form engineered tissues. Reprinted from [3] with permission from IOP Publishing Ltd.

lack of detailed understanding of the interaction between cells and scaffolds. The mechanical properties and surface characteristics affect the behaviour of cells in terms of viability, cell growth, and differentiation [3]. As for the methods used to fabricate bioengineered skin equivalents, three main 3D bioprinting techniques are used: laser-assisted bioprinting (LaBP), (ii) inkjet-based, and (iii) microextrusion-based printing [3, 45–47].

Bioprinting approaches are becoming increasingly popular for various skin TE applications for several reasons. They allow the stiffness and geometry of the hydrogel to be adjusted to accurately mimic the shape of the skin lesion [111]. In addition, the ability to develop custom bioinks can provide optimal printing characteristics while ensuring cell compatibility. Regardless of the specific printing technique, 3D printing allows biomaterials with different mechanical properties to be printed together to increase the scaffold's stability [46, 47]. Moreover, the integration of different 3D bioprinting techniques can mitigate the limitations of each bioprinting technique [3, 45]. To construct a multilayer skin equivalent with a printable, functional transwell system skin, Kim et al. [112] introduced a hybrid bioprinting system, also referred to as an integrated composite tissue or organ building system. They combined extrusion-based and inkjet-based bioprinting simultaneously, using the former to produce a layer similar to the dermis, while the epidermis equivalent was produced using an inkjet module. To further increase the complexity of the 3D-printed skin, they integrated the hypodermis and vascular channels. Based on their porous inlet/outlet transwell platform, they printed a PCL transwell chamber with sequential layers of hypodermis (using adipose tissue-derived decellularized ECM (dECM)-fibrinogen bioink), vascular channels using sacrificial bioinks (human umbilical vein endothelial cells, HUVECs, and thrombin-embedded gelatin hydrogel-based biotin) and dermis, resulting in an artificial vascularized skin construct [6, 45, 112]. The vascular channels supported the underlying hypodermis while allowing for functional interaction with the dermis. An improved epidermal stratification was observed compared to a bilayered scaffold without hypodermis or vascularization. The authors demonstrated that such a full-thickness vascularized skin construct closely mimics natural skin physiology [112]. LaBP-based systems have also been shown to generate bilayer (epidermal-dermal) scaffolds with high spatial resolution and organise sequential layers of fibroblasts and keratinocytes with their respective ECM compositions [6, 113]. Koch and his group were the first to use the LaBP technique to deposit vital cells in a 3D arrangement [113]. They produced stratified layers of fibroblasts and keratinocytes embedded in a collagen hydrogel to demonstrate that the laser-printed cells could form real tissue. These layered sheets were printed onto a sheet of Matriderm™ to demonstrate clinical transferability after printing. In addition, the basal lamina structure, which is a sign of skin tissue formation, in their laser-printed 3D skin model indicated the existence of cell–cell junctions (adherence and gap junctions). Adherence junctions ensure tissue cohesion, while gap junctions enable the direct exchange of metabolites, hormones, electrical signals and messenger substances required for physiological activities [6, 113]. Although replication of the dermal and epidermal layers is important for a successful skin graft, it is not sufficient to mimic native skin's structural, mechanical, and biochemical properties [6, 46, 112].

To model the physiological functions of the skin, it is essential to include the hypodermis and macrostructures such as the vasculature and skin appendages [46, 70]. Several strategies can be followed in the skin TE to promote artificial skin equivalent vascularisation. The classical approach uses sacrificial polymers such as gelatin or agarose printed into a bulk hydrogel like gelatin methacrylate (GelMA) or poly(ethylene glycol diacrylate) (PEGDA) to be later removed from the main matrix. This is followed by perfusion of the produced channels in a subtractive manufacturing process. Establishing an adequate and durable blood supply is essential for the uptake of any tissue construct [46, 114–119]. Therefore, successfully integrating blood vessels into a 3D skin construct is a critical threshold for translation into clinical applications. In other words, to transform an in vitro conceptualized perfused skin construct into a fully vascularized tissue in vivo [114, 116]. In the past, endothelial cells were combined with human dermal fibroblasts to construct a 3D dermal compartment that was stimulated to form capillary-like structures spontaneously. However, in this approach, the formation of vascular networks was random and inefficient developed [112, 120]. Early attempts at forming vascularized 3D skin constructs included bioengineered dermo-epidermal skin grafts with blood and lymphatic capillaries [46, 121]. The pre-vascularized dermo-epidermal skin substitutes were developed using human dermal microvascular endothelial cells (HuDMECs) derived from human foreskin embedded in collagen- or fibrin-based hydrogels. When seeded in 3D hydrogels, a mixture of lymphatic endothelial cells (hLECs) and blood vessel endothelial cells (hBECs) spontaneously formed both lymphatic and blood vessels in the same ECM. Although the researchers demonstrated a milestone in developing the next generation of skin grafts, the blood and lymphatic vascular networks showed incomplete maturation [121]. In addition, InMed Pharmaceuticals, Inc. (Vancouver, Canada) and the TE company ATERA SAS (France) have jointly developed a commercially available vascularized skin model, Skin-VaSc-TERM®. However, further upgrades to it are still being developed [46]. Numerous challenges, including tailoring the vessel diameter, reproducibility, adaptability, and establishing appropriate perfusion culture conditions, limit a rapid and effective vascularization process. In this context, more advanced 3D bioprinting technologies may provide a means to generate a highly organized capillary-sized (1–2 μm) microvasculature with well-defined hierarchical and branching patterns. Endothelial and parenchymal cells, soluble factors and phase-changing hydrogels can be nowadays assembled in a high throughput manner and with high reproducibility [46, 57, 114, 122]. This allows for interlaboratory benchmark that can be used for further comparisons for in vitro therapies and cosmetic studies, and wound dressing evaluation [46]. Having this in mind, Michel et al. [123] used the LaBP technique to create a multi-layered, fully cellularized skin equivalent for the treatment of burn patients. Layers of collagen loaded with fibroblasts and keratinocytes were printed onto a sheet of Matriderm™, which was used as a stabilization matrix. After implantation of these skin constructs into mouse dorsal skinfold chambers, the authors observed the ingrowth of new microvessels from the host tissue. This was likely driven by the seeded keratinocytes' secretion of vascular endothelial growth factor (VEGF) [114, 123]. Further improvement in the construction of vascularized 3D skin analogues was made by Kim and co-workers, who designed an in vitro skin model (described above) that could effectively mimic the skin microcirculation [112]. Briefly, they printed perfusable vascular channels inside the dECM matrix (Fig. 3.6) using HUVECs embedded in the vascular bioink. The latter contained thrombin, which likely cross-linked the dECM fibrinogen bioink to some extent during incubation at 37 °C, indicating its potential use to create blood-perfused vascular channels with minimal structural collapse.

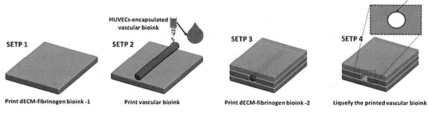

Fig. 3.6 Schematic representation of conceptual fabrication process for forming perfusable vascularized channels. Reproduced from [112] with permission from IOP Publishing Ltd.

Liu et al. developed a technique for versatile and robust bioprinting of skin tissue with increased cellular complexity [124]. The 3D bioprinted skin model was fabricated by combining two advanced TE strategies, 3D bioprinting and electrospinning. Moreover, several cell types, including self-assembled blood vessels, were integrated into the dermis compartment. To do so, the researchers developed a fibrin-based bioink with an embedded mixture of cell types, including neonatal fibroblasts, pericytes, and human-induced pluripotent stem cells (iPSCs)-derived endothelial cells. This bioink loaded with different cell types served as the basis for forming a vascularized dermis printed in a defined spatial pattern on an electrospun PLGA scaffold. In this way, they achieved a reproducible complex 3D cellular culture that enabled high-throughput pharmacological screening in a genuinely physiologically relevant 3D tissue microenvironment [124].

3.3.3.2 Electrospinning

Electrospinning is a versatile and well-established manufacturing method that can produce constructs in the form of non-woven fibre networks [106, 125, 126]. In skin TE, electrospun nanofibers are particularly attractive due to their refined morphology provided by micro- and nanodimensions and their processing flexibility that ultimately allows the formation of unique materials and structures. Together with their ECM-like biomimetic architecture, electrospun nanofibers can promote enhanced re-epithelialization and neotissue formation of wounds [125]. In addition, the electrospun morphology may facilitate the absorption of exudates and allow efficient gas exchange at the wound bed while modulating dermal cell adaptation and proliferation [126, 127]. The process of electrospinning is based on an electrically charged polymer solution that is forced to pass through a nozzle and is affected by an electric field [104, 106]. By applying high voltages, the liquid surface tension of the polymer solution can be overcome, promoting the formation of a polymer jet leading to the formation of a fibrous membrane on a collector [106, 107]. The formation of fibres and their architectural features can be influenced by:

(i)   the properties of the precursor solution (polymer concentration, viscosity, conductivity, surface tension, and solvent),
(ii) the parameters of the electrospinning process (applied voltage, flow rate, tip-to-target distance, collector geometry, needle gauge), and
(iii) ambient conditions (temperature and humidity) [106, 107, 128].

In general, electrospun nanofibers can mimic the ECM's architecture due to their highly interconnected porous structures and high surface-to-volume ratio, which can improve cellular behaviour. Moreover, the high surface-to-volume ratio of electrospun fibres enhances the fluid absorption, dermal drug delivery and shows better absorption of functional proteins such as albumin, fibronectin, and laminin on the surface [106, 107, 129]. The high porosity of nanofibrous scaffolds also facilitates the exchange of oxygen, water, and nutrients, as well as the removal of metabolic wastes. At the same time, small pores limit the penetration of microorganisms [106, 107].

Several approaches incorporate bioactive or therapeutic agents into the electrospun nanofibers for biofunctionalization and drug delivery, including blend, coaxial, and emulsion electrospinning. Just briefly, in blending, the biomolecules are mixed with the polymer solution before the electrospinning process. Since most biomolecules are charged, they migrate towards the surface of the polymer jet due to charge repulsion during electrospinning [106, 130]. In a dual coaxial needle setup, two concentrically arranged nozzles (an outer needle and an inner needle) are used for different solutions to preserve the biomolecules and enable their sustained release. This method creates a core–shell structure where the shell is made of a polymer of choice, and the core is made of encapsulated bioactive agents [106, 107]. The core–shell nanofiber structure is usually associated with effective control of drug release by encapsulating the drug in the core part to prevent burst release, ensure more sustained release and protect the drug from moisture [107]. As the name indicates, emulsion electrospinning uses an emulsion of biomolecules and polymers to embed the bioactive agents into the polymer fibres [106].

For skin regeneration applications, various natural and synthetic polymers and their combinations have been studied to prepare electrospun nanofiber matrices in the form of wound dressings and skin substitutes [104, 106, 111, 131–133]. Duan and colleagues [131] fabricated an electrospun membrane from modified gelatin and PCL (gelatin/PCL) as a carriers used in epidermis engineering. By utilizing the properties of both materials, the electrospun nanofiber gelatin/PCL membrane met the requirements for epidermis engineering in terms of good biocompatibility and sufficient mechanical strength. The fibre network improved permeability, which allowed easy passage of fluids, resulting in improved ability to repair wounds with inflammation and exudates. For this purpose, the authors used two types of cells, immortalized human keratinocytes (HaCaT) and cultured human keratinocytes (KCs) from foreskin samples that were seeded onto the electrospun gelatin/PCL membrane. In vivo analysis revealed that wounds treated with as-prepared epidermis were characterized by accelerated wound closure with limited host response [131]. Keirouz and his group [126] presented trinary biocompatible composite fibres that could be used as starting biomaterials to develop artificial skin constructs with adjustable wettability and enhanced cellular viability. By designing electrospun mats of PCL, silk fibroin (SF) and poly(glycerol sebacate) (PGS) using nozzle-free electrospinning technology, they were able to overcome the shortcomings of pure PCL nanofiber scaffolds. PCL is a popular biomaterial in TE applications due to its good mechanical properties and its ability to imitate many aspects of the native ECM microenvironment. However, its hydrophobic nature limits its applicability. The formed trinary PCL backbone SF/PGS nanofibers exhibited a smooth morphology with controllable fibre diameter and suitable pore size for fibroblast infiltration. By altering the surface properties of the electrospun mats, the PCL-backbone SF/PGS composite material showed good in vitro fibroblast attachment behaviour and optimal growth (Fig. 3.7), indicating the potential of the proposed scaffolds for the development of an artificial skin-like platform [134].

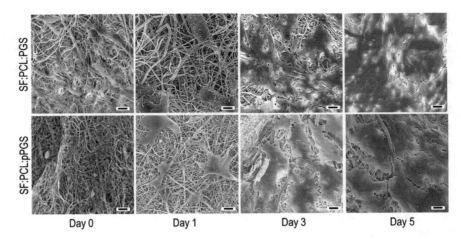

Fig. 3.7 Attachment and proliferation of human dermal fibroblasts (HDFs) on the trinary electrospun membranes. SEM images of HDFs seeded morphology on the trinary SF:PCL electrospun scaffolds. Scale bar is 10 μm. Reproduced from [126] with permission from Elsevier B.V

Electrospun fibrous structures require lengthy incubation times because manual cell seeding is not uniform, and the cells do not fully infiltrate the entire depth of electrospun scaffolds [101, 134, 135]. A novel approach was investigated to address the insufficient infiltration of cells into electrospun scaffolds, which arises from relatively small pores of such materials. The cell-polymer solution was spun in a single step, called "cell electrospinning". Townsend-Nicholson and Jayasinghe [115] fabricated scaffolds in tandem consisting of multiple cell types using this unique technique. They used coaxial electrospinning in which a living bio-suspension flowed through the inner needle. In contrast, a low conducting high viscosity polymer (medical grade poly(dimethylsiloxane), PDMS, medium) flowed through the outer needle. In this manufacturing setup, the rotating conducting collector was immersed in a cell culture medium containing essential nutrients to prevent cell dehydration [136]. In terms of survival of the treated cells, electrospinning of cells can be considered as a safe method to develop 3D biological models. Moreover, even after transfer, the living composite structures integrated into their hosts without causing any kind of rejection [137].

While mechanical collectors for template-assisted 3D electrospinning are desirable for certain TE applications, particularly those requiring tubular structures, they are less commonly used in skin TE [125]. Customized electrospinning collectors have been used to control the pore size of 3D structures by providing a spherical template with an array of point collectors [138]. Moreover, the combination with 3D bioprinted scaffolds serving as templates facilitates the construction of electrospun platforms with tunable pore size and mechanical properties. In this context, Miguel et al. [111] used a 3D bioprinted CS/ALG hydrogel onto which they electrospun a mixture of PCL and silk sericin (PCL_SS) to recreate the complex features of the epidermis. The asymmetric 3D nanofiber construct of the epidermis showed comparable mechanical properties to natural skin and good antimicrobial activity [111]. Hadar and colleagues [139] went one step further and used various fabrication techniques, including casting, electrospinning, and lyophilization, to design a three-layer regenerative skin scaffold that can be applied to the wound site in a single step. The developed three-layer PCL gelatin scaffold was similar to actual skin layers (epidermis, dermis, and hypodermis) in terms of architecture and physical and mechanical properties. The in vitro co-cultured model prepared with keratinocytes and fibroblasts suggests that a multilayered scaffold can keep each cell population separate while promoting their respective proliferation and ECM deposition. Moreover, the in vivo results confirmed the efficiency of wound healing, and the regenerated tissue showed morphological similarity to native skin [139].

Some 3D electrospinning techniques are commonly used for various TE applications. The resulting 3D nanofiber structures are also beneficial in skin TE, especially in relation to better control over pore size and overall scaffold porosity. Self-assembly of 3D fibrous materials by electrospinning is an emerging area of research as desirable structures can be fabricated in a single step for use as TE scaffolds. Progress has been limited specifically for application to skin TE, although further advancement in this area holds promise for the development of 3D nanofiber scaffolds characterized by improved morphological control and a greater variety of materials for improved skin TE [125].

3.3.4 Skin-On-Chip

Technological developments at the interface of microfluidics and cell culture have led to microphysiological systems that aim to circumvent the disabilities of conventional 3D cell culture platforms cultured under static conditions. By modelling critical functional units of organs in microfluidic platforms, the aim is to reproduce the biological composition, function and environment of native tissues and bridge the gap between standard in vitro approaches and the human body [33, 115, 140]. The organ-on-a-chip technology is superior to traditional 3D cultures in two ways:

(i) transport of substances is more physiologically relevant, allowing for a more realistic assessment of parameters,
(ii) microfluidics maintains the high throughput capacity of the systems while reducing costs and quantities of reagents needed [115].

Skin-on-chip (SoC) devices have proven their merit in improving the barrier function of skin equivalents, thickening the epidermis, and promoting keratinocyte differentiation. Because SoC models better mimic human tissues, they are ideal candidates for drug screening and the discovery and/or repositioning of pharmacologically relevant molecules. These microfluidic devices allow tissue culturing to control various physical and biochemical parameters, such as flows, forces, or chemical gradients. Typically, SoCs contain a transwell holder or porous membrane that separates either the epidermis from the dermis and epithelial cells or the engineered skin from the underlying perfused culture medium to allow macromolecules to diffuse into the skin model. Different strategies for developing a SoC have been presented, differing greatly in key aspects such as the manufacturing process and materials or tissue maintenance [33, 115]. In essence, the most common are:

(i) direct introduction of a skin fragment obtained from a biopsy or cultured human skin equivalent (HSE) in vitro into the chip,
(ii) in situ generation of skin directly on a chip.

The first approach, also known as transferred SoC, is common practice in multi-organ chip development [115, 141]. The use of well-formed and mature skin fragments allows for more realistic models as different skin layers are present. Some research groups have already used commercially available skin equivalents such as EpiDerm™ from which they have built SoCs [115, 142, 143]. For example, Ataç et al. used a commercially available bilayer skin substitute EpiDermFT™ on which they modelled a "hairy" SoC by incorporating complete hair follicle units and hypodermis from human biopsies [143]. The presence of hypodermis prolonged the viability of the in vitro skin equivalent (EpiDermFT™) in both static and SoC cultures with improved preservation of tissue architecture. The proposed SoC system with dynamic perfusion also prolonged the hair follicle viability and postponed the onset of the catagen phase of hair growth (end of active growth) [115, 143]. These skin models do not completely conform with the original definition of organ-on-a-chip, i.e. a microfluidic cell culture device created using microchip fabrication methods containing continuously perfused chambers inhabited by living cells arranged to simulate tissue- and organ-level physiology [144]. Nevertheless, they have enabled studies of several factors affecting the maintenance of skin substitutes and shed more light on their use for clinical and testing purposes (molecular diffusion, multi-organ crosstalk, drug sensitivity, and toxicity) [115].

The second strategy, in situ generation, focuses on fabricating the skin model directly on the chip, which can be achieved in two ways. The first is based on an artificially vascularized dermis manually generated in an open structure within the device. In principle, it is similar to the transferred SoC systems, but the main difference lies in how the skin constructs are supplied with culture medium or other substances. The transferred SoCs are perfused through hollow channels passing across the dermal compartment. In contrast, the circulation of fluids in the in situ SoCs occurs through a microfluidic channel below the tissue construct [115, 145–149]. In the second approach of in situ modelling of SoC, microfluidic channels serve as a pathway for the delivery of nutrients and as anchors for tissue support. The use of channels as compartments for culturing skin tissue may have some disadvantages related to the complexity of replicating the 3D architecture of the natural tissue. These drawbacks can be easily overcome by designing a device with a well-like structure and channels separated by a porous membrane [115, 150–152]. With this setup, the dermo-epidermal structure can be mimicked. Moreover, the introduction of microfluidic cultures can improve the barrier function of the tissue and promote the synthesis of basement membrane proteins [150]. The first vascularized in situ SoC (Fig. 3.8),

which more closely resembles the actual architecture of native skin, was designed by
Wufuer et al. [152]. Although the authors used cell monolayers, the system consisted
of three channels to mimic the epidermis, dermis, and blood vessels, separated by a
porous membrane. Immortalized keratinocytes (HaCaTs) were seeded on the upper
membrane and HUVECs on the bottom of the lower membrane to simulate the
epidermis and the endothelium covering the vessels. The dermal compartment of the
SoC was modelled through the formation of fibroblast monolayers on the bottom of
the upper membrane and the top of the lower membrane, with the middle channel
[152].

In microfluidic organ-on-a-chip systems, a porous membrane is typically used
to separate the skin equivalent from the medium to allow oxygen and nutrients
to diffuse from the medium into the dermal layer. Many SoC systems only facilitate this single-sided medium perfusion on the basal layer with dermis, while the
epidermis is often not exposed. Fluid flow-induced shear stress is known to be a
mechanical stimulus for several epithelial cell types, including keratinocytes. Exposure of keratinocytes to low shear stress affected the structural reorganization of
the epidermis model, resulting in an improved epithelial phenotype and a thicker
viable epidermis [153]. In addition, it has been demonstrated that other mechanical

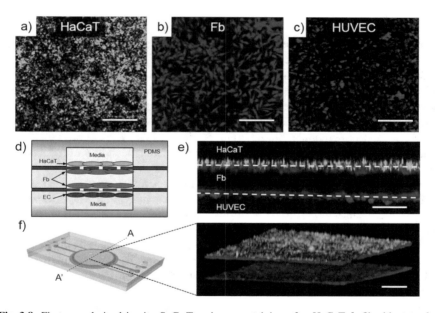

Fig. 3.8 First vascularized in situ SoC. Top: immunostaining of **a** HaCaT, **b** fibroblasts and **c**
HUVEC. In the middle: **d** illustration of the side view of SoC, and **e** Z-stacked fluorescence image
showing all three cell types adhered on two porous membranes. At the bottom: **f** 3D fluorescence
image of a cross section of the SoC. Scale bars: 300 μm. Reproduced from [152] with permission
from Springer Nature Limited

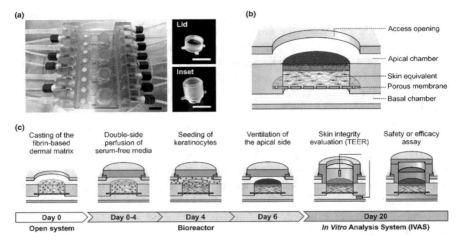

Fig. 3.9 Microfluidic skin equivalent (full-thickness SoC). **a** images of the assembled SoC and its bioreactor setup, together with interchangeable lid and inset, **b** graphic abstract of the full-thickness SoC for coculture of human skin fibroblasts and keratinocytes, and **c** an overview of the full-thickness SoC and in vitro assay protocols. Reproduced from [150] with permission from Elsevier B.V

stimuli (uniaxial/biaxial stretching, matrix stiffness, cyclic loading) can also influence the physiological aspects of keratinocytes [154–157]. The positive effect of bilateral perfusion on skin barrier function in an SoC system was demonstrated by Sriram and colleagues [150]. They developed a human skin equivalent (Fig. 3.9), which combined a fibrin-based dermal matrix with a dynamic perfusion microfluidic platform. The latter was implemented using a peristaltic pump that allowed the culture medium to circulate through microfluidic compartments ensuring a continuous supply of nutrients and simultaneous removal of metabolic waste products, thus functioning similarly to the blood vessels in natural human tissues [150].

Another novelty of this SoC system was using a fibrin-based dermal matrix, which, due to its noncontractile properties, allows integrated 3D cultures and integrity/permeability tests to be performed directly on the device [33, 150]. Like other in vitro skin models, the most commonly utilized ECM material in SoC platforms is collagen, which is known to shrink and detach from the membrane during fibroblast proliferation [33, 158, 159]. Therefore, the introduction of fibrin instead of collagen as an ECM component has alleviated the limitations of collagen-based skin equivalents used in conventional cell culture inserts and diffusion cells [33, 150]. In addition, histological examination showed that the skin equivalents exposed to shear stress more closely resembled natural skin tissue (Fig. 3.10).

Microfluidics, with their microscale size, offers fine control of the fluid environment and coherent application of shear stress to both basal and apical layers (as illustrated in Fig. 3.9). This has strong implications for improving the quality and biomimicry of bioengineered skin models [33, 150].

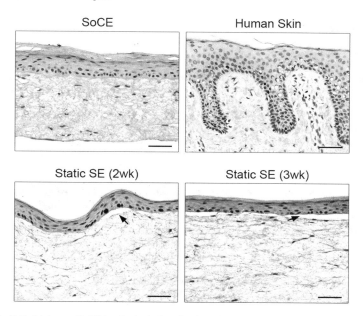

Fig. 3.10 Full-thickness SoC histological visualization: the impact of microfluidics on epidermal homeostasis. Representative photomicrographs of immunoperoxidase staining for proliferation marker, Ki67 (brown nuclear staining). Black arrows point to the artefactual separation of the epidermis from the underlying dermis, which commonly occurs in static skin equivalents (SEs). Reproduced from [150] with permission from Elsevier B.V

Although medium perfusion is key in SoC systems, the introduction of perfusable blood vessels is relatively new, and protocols have yet to be optimized. Vascularization strategies can be distinguished between pre-vascularization, in which microvessel channels are made prior to endothelial cell inoculation, or angiogenesis, in which GFs stimulate endothelial cells to vascularize the tissue. Angiogenesis is a slow process, unsuitable for larger tissue constructs or efficient high-throughput applications. Therefore, methods for pre-vascularization are generally preferred [33, 160]. Blood vessels modelled in microfluidic skin equivalents have not reached dimensions relevant to capillaries, as seen in alternative skin models fabricated in an angiogenic manner [33, 161]. Combined with recent advances in 3D bioprinting techniques, vascularized SoC systems offer the possibility of automated and realistic high-throughput testing, applicable in regenerative medicine, cosmetics, and pharmaceutical applications [33]. To date, great progress has been made in developing a biomimetic skin analogue using SoC devices. However, the lack of understanding of the complex synergistic effects between epidermal keratinocytes and dermal fibroblasts is delaying the development of a device that fully replicates the critical features of native skin tissue and represents a truly complete approach that can replace animal testing [33].

References

1. Sanabria-de la Torre R, Fernández-González AFV, Quiñones-Vico MI, Montero-Vilchez T, Arias-Santiago S (2020) Bioengineered skin intended as in vitro model for pharmacosmetics, skin disease XE "skin disease" study and environmental skin impact analysis. Biomedicines 8(11):464

2. Oualla-Bachiri W, Fernández-González A, Quiñones-Vico MI, Arias-Santiago S (2020) From grafts to human bioengineered vascularized skin substitutes. Int J Mol Sci 21(21):8197

3. Vijayavenkataraman S, Lu W, Fuh J (2016) 3D bioprinting XE "3D bioprinting" of skin: a state-of-the-art review on modelling, materials, and processes. Biofabrication 8(3):032001

4. Low ZWK, Li Z, Owh C, Chee PL, Ye E, Dan K, Chan SY, Young DJ, Loh XJ (2020) Recent innovations in artificial skin. Biomaterials science 8(3):776–797

5. Dolbashid AS, Mokhtar MS, Muhamad F, Ibrahim F (2017) Potential applications of human artificial skin and electronic skin (e-skin): a review. Bioinspired, Biomimetic Nanobiomaterials 7(1):53–64

6. Yu JR, Navarro J, Coburn JC, Mahadik B, Molnar J, Holmes JH IV, Nam AJ, Fisher JP (2019) Current and future perspectives on skin tissue engineering: key features of biomedical research, translational assessment, and clinical application. Adv Healthcare Mater 8(5):1801471

7. Spear M (2011) Skin grafts: indications. Appl Current Res (2011)

8. Thornton JF, Gosman A (2004) Skin grafts and skin substitutes. Sel Readings Plast Surg 10(1):1–24

9. Cahn B, Lev-Tov H (2020) Cellular-and acellular-based therapies: skin substitutes and matrices. In: Alavi A, Maibach H (eds) Local wound care for dermatologists. Updates in clinical dermatology, pp 139–151

10. Erovic BM, Lercher P (2015) Skin graft harvesting, manual of head and neck reconstruction using regional and free flaps. Springer Vienna, Vienna, pp 53–60

11. Bennett J (1983) Aspects of the history of plastic surgery since the 16th century. J R Soc Med 76(2):152–156

12. Fairbairn N, Randolph M, Redmond R (2014) The clinical applications of human amnion XE "amnion" in plastic surgery. J Plastic Reconstr Aesthetic Surg 67(5):662–675

13. Karasek MA (1968) Growth and differentiation of transplanted epithelial cell cultures. J Invest Dermatol 51(4):247–252

14. Rheinwatd JG, Green H (1975) Seria cultivation of strains of human epidemal keratinocytes XE "keratinocytes": the formation keratinizin colonies from single cell is. Cell 6(3):331–343

15. O'Connor N, Mulliken J, Banks-Schlegel S, Kehinde O, Green H (1981) Grafting of burns XE "burns" with cultured epithelium prepared from autologous epidermal cells. The Lancet 317(8211):75–78

16. Compton C, Gill J, Bradford D, Regauer S, Gallico G, O'connor N (1989) Skin regenerated from cultured epithelial autografts on full-thickness burn wounds from 6 days to 5 years after grafting. A light, electron microscopic and immunohistochemical study. Lab Invest J Tech Methods Pathol 60(5):600–612

17. Green H, Kehinde O, Thomas J (1979) Growth of cultured human epidermal cells into multiple epithelia suitable for grafting. Proc Natl Acad Sci 76(11):5665–5668

18. Burke JF, Yannas IV, Quinby WC Jr, Bondoc CC, Jung WK (1981) Successful use of a physiologically acceptable artificial skin in the treatment of extensive burn injury. Ann Surg 194(4):413

19. Yannas I, Burke J, Orgill D, Skrabut E (1982) Wound tissue can utilize a polymeric template to synthesize a functional extension of skin. Science 215(4529):174–176

20. Eaglstein WH, Falanga V (1997) Tissue engineering and the development of Apligraf® XE "Apligraf®", a human skin equivalent. Clin Ther 19(5):894–905

21. Phillips T, Gilchrest B (1992) Clinical applications of cultured epithelium. Epithelial Cell Biol 1(1):39–46

22. Moiemen NS, Vlachou E, Staiano JJ, Thawy Y, Frame JD (2006) Reconstructive surgery with Integra dermal regeneration template: histologic study, clinical evaluation, and current practice. Plastic Reconstr Surg 117(7S):160S-174S

23. Chang DK, Louis MR, Gimenez A, Reece EM (2019) The basics of integra dermal regeneration template and its expanding clinical applications. In: Seminars in plastic surgery. Thieme Medical Publishers, pp 185–189

24. Doillon CJ, Wasserman AJ, Berg RA, Silver FH (1988) Behaviour of fibroblasts XE "fibroblasts" and epidermal cells cultivated on analogues of extracellular matrix. Biomaterials 9(1):91–96

25. Kun M, Chan C, Ramakrishna S, Kulkarni A, Vadodaria K (2019) Textile-based scaffolds for tissue engineering. Elsevier, Advanced textiles for wound care, pp 329–362

26. Cuono C, Langdon R, McGuire J (1986) Use of cultured epidermal autografts and dermal allografts XE "allograft" as skin replacement after burn injury. The Lancet 327(8490):1123–1124

27. Langer R, Vacanti JP (1993) Tissue engineering. Science 260(5110):920–927

28. Luu-The V, Duche D, Ferraris C, Meunier J-R, Leclaire J, Labrie F (2009) Expression profiles of phases 1 and 2 metabolizing enzymes in human skin and the reconstructed skin models Episkin™ and full thickness model from Episkin™. J Steroid Biochem Mol Biol 116(3–5):178–186

29. Netzlaff F, Lehr C-M, Wertz P, Schaefer U (2005) The human epidermis XE "epidermis" models EpiSkin®, SkinEthic® and EpiDerm®: An evaluation of morphology and their suitability for testing phototoxicity, irritancy, corrosivity, and substance transport. Eur J Pharm Biopharm 60(2):167–178

30. Eilstein J, Léreaux G, Arbey E, Daronnat E, Wilkinson S, Duché D (2015) Xenobiotic metabolizing enzymes in human skin and SkinEthic reconstructed human skin models. Exp Dermatol 24(7):547–549

31. Pageon H, Zucchi H, Dai Z, Sell DR, Strauch CM, Monnier VM, Asselineau D (2015) Biological effects induced by specific advanced glycation end products in the reconstructed skin model of aging. BioResearch Open Access 4(1):54–64

32. Duval K, Grover H, Han LH, Mou Y, Pegoraro AF, Fredberg J, Chen Z (2017) Modeling physiological events in 2D versus 3D cell culture. Physiology 32(4):266–277

33. Sutterby E, Thurgood P, Baratchi S, Khoshmanesh K, Pirogova E (2020) Microfluidic skin-on-a-chip models: toward biomimetic artificial skin. Small 16(39):2002515

34. Edmondson R, Broglie JJ, Adcock AF, Yang L (2014) Three-dimensional cell culture systems and their applications in drug discovery and cell-based biosensors. Assay Drug Dev Technol 12(4):207–218

35. Bäsler K, Bergmann S, Heisig M, Naegel A, Zorn-Kruppa M, Brandner JM (2016) The role of tight junctions in skin barrier XE "barrier" function and dermal absorption. J Control Release 242:105–118

36. Ihalainen TO, Aires L, Herzog FA, Schwartlander R, Moeller J, Vogel V (2015) Differential basal-to-apical accessibility of lamin A/C epitopes in the nuclear lamina regulated by changes in cytoskeletal tension. Nat Mater 14(12):1252–1261

37. Huh D, Kim HJ, Fraser JP, Shea DE, Khan M, Bahinski A, Hamilton GA, Ingber DE (2013) Microfabrication of human organs-on-chips. Nat Protoc 8(11):2135–2157

38. Chaubey A, Ross KJ, Leadbetter RM, Burg KJ (2008) Surface patterning: tool to modulate stem cell differentiation in an adipose system. J Biomed Mater Res B Appl Biomater 84(1):70–78

39. Ballester-Beltrán J, Lebourg M, Salmerón-Sánchez M (2015) Sandwich-like microenvironments to harness cell/material interactions. J Visualized Exp: JoVE (102)

40. Ballester-Beltrán J, Lebourg M, Salmerón-Sánchez M (2013) Dorsal and ventral stimuli in sandwich-like microenvironments. Effect on cell differentiation, Biotechnology Bioengineering 110(11):3048–3058

41. Dunn J, Tompkins RG, Yarmush ML (1992) Hepatocytes in collagen XE "collagen" sandwich: evidence for transcriptional and translational regulation. J Cell Biol 116(4):1043–1053

42. Ezzell RM, Toner M, Hendricks K, Dunn JC, Tompkins RG, Yarmush ML (1993) Effect of collagen XE "collagen" gel configuration on the cytoskeleton in cultured rat hepatocytes. Exp Cell Res 208(2):442–452
43. Jones HM, Barton HA, Lai Y, Bi Y-A, Kimoto E, Kempshall S, Tate SC, El-Kattan A, Houston JB, Galetin A (2012) Mechanistic pharmacokinetic modeling for the prediction of transporter-mediated disposition in humans from sandwich culture human hepatocyte data. Drug Metab Dispos 40(5):1007–1017
44. Garcia M, Escamez MJ, Carretero M, Mirones I, Martinez-Santamaria L, Navarro M, Jorcano JL, Meana A, Del Rio M, Larcher F (2007) Modeling normal and pathological processes through skin tissue engineering. Mol Carcinog Published in cooperation with the University of Texas MD Anderson Cancer Center 46(8):741–745
45. Choudhury S, Das A (2020) Advances in generation of three-dimensional skin equivalents: pre-clinical studies to clinical therapies. Cytotherapy
46. Randall MJ, Jüngel A, Rimann M, Wuertz-Kozak K (2018) Advances in the Biofabrication of 3D Skin *in vitro*: Healthy and Pathological Models. Front Bioeng Biotechnol 6:154
47. Sarkiri M, Fox SC, Fratila-Apachitei LE, Zadpoor AA (2019) Bioengineered skin intended for skin disease XE "skin disease" modeling. Int J Mol Sci 20(6):1407
48. Antoni D, Burckel H, Josset E, Noel G (2015) Three-dimensional cell culture: a breakthrough *in vivo*. Int J Mol Sci 16(3):5517–5527
49. Langhans SA (2018) Three-dimensional in vitro cell culture models in drug discovery and drug repositioning. Front Pharmacol 9:6
50. Ovsianikov A, Khademhosseini A, Mironov V (2018) The synergy of scaffold-based and scaffold-free tissue engineering strategies. Trends Biotechnol 36(4):348–357
51. Alghuwainem A, Alshareeda AT, Alsowayan B (2019) Scaffold-free 3-D cell sheet technique bridges the gap between 2-D cell culture and animal models. Int J Mol Sci 20(19):4926
52. Kinikoglu B (2017) A comparison of scaffold-free and scaffold-based reconstructed human skin models as alternatives to animal use. Altern Lab Anim 45(6):309–316
53. Białkowska K, Komorowski P, Bryszewska M, Miłowska K (2020) Spheroids as a type of three-dimensional cell cultures—Examples of methods of preparation and the most important application. Int J Mol Sci 21(17):6225
54. Foty R (2011) A simple hanging drop cell culture protocol for generation of 3D spheroids. J Visualized Exp: JoVE (51)
55. Ströbel S, Buschmann N, Neeladkandhan A, Messner S, Kelm J (2016) Characterization of a novel *in vitro* 3D skin microtissue model for efficacy and toxicity testing. Toxicol Lett (258):S156–S157
56. Klicks J, Maßlo C, Kluth A, Rudolf R, Hafner M (2019) A novel spheroid-based co-culture model mimics loss of keratinocyte XE "keratinocyte" differentiation, melanoma cell invasion, and drug-induced selection of ABCB5-expressing cells. BMC Cancer 19(1):1–14
57. Liu Y, Luo H, Wang X, Takemura A, Fang YR, Jin Y, Suwa F (2013) In vitro construction of scaffold-free bilayered tissue-engineered skin containing capillary networks. BioMed Res Int 2013
58. Michel M, L'Heureux N, Pouliot R, Xu W, Auger FA, Germain L (1999) Characterization of a new tissue-engineered human skin equivalent with hair. In vitro Cell Develop Biol-Anim 35(6):318
59. Danilenko DM, Phillips GDL, Diaz D (2016) In vitro skin models and their predictability in defining normal and disease biology, pharmacology, and toxicity. Toxicol Pathol 44(4):555–563
60. Bhardwaj N, Chouhan D, Mandal BB (2018) 3D functional scaffolds for skin tissue engineering. In: Functional 3D tissue engineering scaffolds. Elsevier2018, pp 345–365
61. Vrana NE, Lavalle P, Dokmeci MR, Dehghani F, Ghaemmaghami AM, Khademhosseini A (2013) Engineering functional epithelium for regenerative medicine and *in vitro* organ models: a review. Tissue Eng Part B Rev 19(6):529–543
62. Chan EC, Kuo S-M, Kong AM, Morrison WA, Dusting GJ, Mitchell GM, Lim SY, Liu G-S (2016) Three dimensional collagen XE "collagen" scaffold promotes intrinsic vascularisation for tissue engineering applications. PLoS ONE 11(2):e0149799

63. Mathew-Steiner SS, Roy S, Sen CK (2021) Collagen XE "collagen" in wound healing. Bioengineering 8(5):63

64. Varghese SA, Rangappa SM, Siengchin S, Parameswaranpillai J (2020) Natural polymers and the hydrogels prepared from them. In: Chen Y (ed) Hydrogels based on natural polymers. Elsevier2020, pp 17–47

65. Catoira MC, Fusaro L, Di Francesco D, Ramella M, Boccafoschi F (2019) Overview of natural hydrogels for regenerative medicine applications. J Mater Sci - Mater Med 30(10):1–10

66. Moulin V, Castilloux G, Jean A, Garrel D, Auger FA, Germain L (1996) In vitro models to study wound healing XE "wound healing" fibroblasts XE "fibroblasts." Burns 22(5):359–362

67. Vidal SEL, Tamamoto KA, Nguyen H, Abbott RD, Cairns DM, Kaplan DL (2019) 3D biomaterial matrix to support long term, full thickness, immuno-competent human skin equivalents with nervous system components. Biomaterials 198:194–203

68. Brown RA, Wiseman M, Chuo CB, Cheema U, Nazhat SN (2005) Ultrarapid engineering of biomimetic materials and tissues: Fabrication of nano-and microstructures by plastic compression. Adv Func Mater 15(11):1762–1770

69. Braziulis E, Diezi M, Biedermann T, Pontiggia L, Schmucki M, Hartmann-Fritsch F, Luginbühl J, Schiestl C, Meuli M, Reichmann E (2012) Modified plastic compression of collagen XE "collagen" hydrogels provides an ideal matrix for clinically applicable skin substitutes. Tissue Eng Part C Methods 18(6):464–474

70. Schmidt FF, Nowakowski S, Kluger PJ (2020) Improvement of a three-layered *in vitro* skin model for topical application of irritating substances. Front Bioeng Biotechnol 8:388

71. Llames SG, Del Rio M, Larcher F, García E, García M, Escamez MJ, Jorcano JL, Holguín P, Meana A (2004) Human plasma as a dermal scaffold for the generation of a completely autologous bioengineered skin. Transplantation 77(3):350–355

72. Xu S, Sang L, Zhang Y, Wang X, Li X (2013) Biological evaluation of human hair keratin scaffolds for skin wound repair and regeneration. Mater Sci Eng C 33(2):648–655

73. Park YR, Ju HW, Lee JM, Kim D-K, Lee OJ, Moon BM, Park HJ, Jeong JY, Yeon YK, Park CH (2016) Three-dimensional electrospun silk-fibroin nanofiber for skin tissue engineering. Int J Biol Macromol 93:1567–1574

74. Hartmann-Fritsch F, Biedermann T, Braziulis E, Luginbühl J, Pontiggia L, Böttcher-Haberzeth S, van Kuppevelt TH, Faraj KA, Schiestl C, Meuli M (2016) Collagen XE "collagen" hydrogels strengthened by biodegradable meshes are a basis for dermo-epidermal skin grafts XE "skin grafts" intended to reconstitute human skin in a one-step surgical intervention. J Tissue Eng Regenerative Med 10(1):81–91

75. Sarkar SD, Farrugia BL, Dargaville TR, Dhara S (2013) Chitosan–collagen XE "collagen" scaffolds with nano/microfibrous architecture for skin tissue engineering. J Biomed Mater Res Part A Official J Soc Biomaterials, Jpn Soc Biomaterials, Aust Soc Biomaterials Korean Soc Biomaterials 101(12):3482–3492

76. Sharif S, Ai J, Azami M, Verdi J, Atlasi MA, Shirian S, Samadikuchaksaraei A (2018) Collagen XE "collagen" -coated nano-electrospun PCL XE "PCL" seeded with human endometrial stem cells for skin tissue engineering applications. J Biomed Mater Res B Appl Biomater 106(4):1578–1586

77. Farrugia BL, Brown TD, Upton Z, Hutmacher DW, Dalton PD, Dargaville TR (2013) Dermal fibroblast infiltration of poly (ε-caprolactone) scaffolds fabricated by melt electrospinning XE "electrospinning" in a direct writing mode. Biofabrication 5(2):025001

78. Pal P, Dadhich P, Srivas PK, Das B, Maulik D, Dhara S (2017) Bilayered nanofibrous 3D hierarchy as skin rudiment by emulsion electrospinning XE "electrospinning" for burn wound management. Biomaterials Sci 5(9):1786–1799

79. Suhail S, Sardashti N, Jaiswal D, Rudraiah S, Misra M, Kumbar SG (2019) Engineered skin tissue equivalents for product evaluation and therapeutic applications. Biotechnol J 14(7):1900022

80. Raney SG, Franz TJ, Lehman PA, Lionberger R, Chen M-L (2015) Pharmacokinetics-based approaches for bioequivalence evaluation of topical dermatological drug products. Clin Pharmacokinet 54(11):1095–1106

81. Martínez-Santamaría L, Guerrero-Aspizua S, Del Río M (2012) Skin bioengineering: preclinical and clinical applications. Actas Dermo-Sifiliográficas (English Edition) 103(1):5–11
82. Bataillon M, Lelièvre D, Chapuis A, Thillou F, Autourde JB, Durand S, Boyera N, Rigaudeau A-S, Besné I, Pellevoisin C (2019) Characterization of a new reconstructed full thickness skin model, T-Skin™, and its application for investigations of anti-aging compounds. Int J Mol Sci 20(9):2240
83. Bell E, Sher S, Hull B, Merrill C, Rosen S, Chamson A, Asselineau D, Dubertret L, Lapiere C, Neveux Y (1983) The reconstitution of living skin. J Investig Dermatol 81(1):S2–S10
84. OECD TN, 439 (2015) In vitro skin irritation: reconstructed human epidermis test method. OECD Guidelines Testing Chemicals, Section 4
85. Szymański Ł, Jęderka K, Cios A, Ciepelak M, Lewicka A, Stankiewicz W, Lewicki S (2020) A simple method for the production of human skin equivalent in 3D, multi-cell culture. Int J Mol Sci 21(13):4644
86. Carlson MW, Alt-Holland A, Egles C, Garlick JA (2008) Three-dimensional tissue models of normal and diseased skin. Curr Protoc Cell Biol 41(1):19.9. 1–19.9. 17
87. Reijnders CM, van Lier A, Roffel S, Kramer D, Scheper RJ, Gibbs S (2015) Development of a full-thickness human skin equivalent in vitro model derived from TERT-immortalized keratinocytes XE "keratinocytes" and fibroblasts XE "fibroblasts." Tissue Eng Part A 21(17–18):2448–2459
88. Yan W-C, Davoodi P, Vijayavenkataraman S, Tian Y, Ng WC, Fuh JY, Robinson KS, Wang C-H (2018) 3D bioprinting XE "3D bioprinting" of skin tissue: from pre-processing to final product evaluation. Adv Drug Deliv Rev 132:270–295
89. Pruniéras M, Régnier M, Woodley D (1983) Methods for cultivation of keratinocytes XE "keratinocytes" with an air-liquid interface XE "air-liquid interface." J Investig Dermatol 81(1):S28–S33
90. Mewes K, Raus M, Bernd A, Zöller N, Sättler A, Graf R (2007) Elastin XE "elastin" expression in a newly developed full-thickness skin equivalent. Skin Pharmacol Physiol 20(2):85–95
91. Stellavato A, Corsuto L, D'Agostino A, La Gatta A, Diana P, Bernini P, De Rosa M, Schiraldi C (2016) Hyaluronan hybrid cooperative complexes as a novel frontier for cellular bioprocesses re-activation. PLoS ONE 11(10):e0163510
92. Ackermann K, Borgia SL, Korting HC, Mewes K, Schäfer-Korting M (2010) The Phenion® full-thickness skin model for percutaneous absorption testing. Skin Pharmacol Physiol 23(2):105–112
93. Batheja P, Song Y, Wertz P, Michniak-Kohn B (2009) Effects of growth conditions on the barrier XE "barrier" properties of a human skin equivalent. Pharm Res 26(7):1689–1700
94. Jäckh C, Blatz V, Fabian E, Guth K, van Ravenzwaay B, Reisinger K, Landsiedel R (2011) Characterization of enzyme activities of Cytochrome P450 enzymes, Flavin-dependent monooxygenases, N-acetyltransferases and UDP-glucuronyltransferases in human reconstructed epidermis XE "epidermis" and full-thickness skin models. Toxicol In Vitro 25(6):1209–1214
95. Reisinger K, Blatz V, Brinkmann J, Downs TR, Fischer A, Henkler F, Hoffmann S, Krul C, Liebsch M, Luch A (2018) Validation of the 3D skin comet assay using full thickness skin models: transferability and reproducibility, mutation research/genetic toxicology and environmental. Mutagenesis 827:27–41
96. Silva R, Ferreira H, Matamá T, Gomes AC, Cavaco-Paulo A (2012) Wound-healing evaluation of entrapped active agents into protein microspheres over cellulosic gauzes. Biotechnol J 7(11):1376–1385
97. Vasconcelos A, Gomes AC, Cavaco-Paulo A (2012) Novel silk fibroin/elastin wound dressings. Acta Biomater 8(8):3049–3060
98. Kuchler S, Wolf N, Heilmann S, Weindl G, Helfmann J, Yahya M, Stein C, Schafer-Korting M (2010) 3D-wound healing model: influence of morphine and solid lipid nanoparticles. J Biotechnol 148
99. Orlov MD, Chernyavsky AI, Arredondo J, Grando SA (2006) Synergistic actions of pemphigus vulgaris IgG, Fas-ligand and tumor necrosis factor-α during induction of basal cell shrinkage and acantholysis. Autoimmunity 39(7):557–562

100. Lee S, Dong DX, Jindal R, Maguire T, Mitra B, Schloss R, Yarmush M (2014) Predicting full thickness skin sensitization using a support vector machine. Toxicol In Vitro 28(8):1413–1423

101. Savoji H, Godau B, Hassani MS, Akbari M (2018) Skin tissue substitutes and biomaterial risk assessment and testing. Front Bioeng Biotechnol 6:86

102. Santschi M, Vernengo A, Eglin D, D'Este M, Wuertz-Kozak K (2019) Decellularized matrix as a building block in bioprinting and electrospinning XE "electrospinning." Curr Opin Biomed Eng 10:116–122

103. Pedde RD, Mirani B, Navaei A, Styan T, Wong S, Mehrali M, Thakur A, Mohtaram NK, Bayati A, Dolatshahi-Pirouz A (2017) Emerging biofabrication strategies for engineering complex tissue constructs. Adv Mater 29(19):1606061

104. Tottoli EM, Dorati R, Genta I, Chiesa E, Pisani S, Conti B (2020) Skin wound healing XE "wound healing" process and new emerging technologies for skin wound care and regeneration. Pharmaceutics 12(8):735

105. Rahmati M, Mills DK, Urbanska AM, Saeb MR, Venugopal JR, Ramakrishna S, Mozafari M (2021) Electrospinning for tissue engineering applications. Prog Mater Sci 117:100721

106. Norouzi M, Boroujeni SM, Omidvarkordshouli N, Soleimani M (2015) Advances in skin regeneration: application of electrospun scaffolds. Adv Healthcare Mater 4(8):1114–1133

107. Arida IA, Ali IH, Nasr M, El-Sherbiny IM (2021) Electrospun polymer-based nanofiber scaffolds for skin regeneration. J Drug Delivery Sci Technol:102623

108. Ramasamy S, Davoodi P, Vijayavenkataraman S, Teoh JH, Thamizhchelvan AM, Robinson KS, Wu B, Fuh JY, DiColandrea T, Zhao H (2021) Optimized construction of a full thickness human skin equivalent using 3D bioprinting XE "3D bioprinting" and a PCL XE "PCL"/collagen XE "collagen" dermal scaffold. Bioprinting 21:e00123

109. Augustine R (2018) Skin bioprinting: a novel approach for creating artificial skin from synthetic and natural building blocks. Prog Biomater 7(2):77–92

110. Pereira RF, Barrias CC, Granja PL, Bartolo PJ (2013) Advanced biofabrication strategies for skin regeneration and repair. Nanomedicine 8(4):603–621

111. Miguel SP, Cabral CS, Moreira AF, Correia IJ (2019) Production and characterization of a novel asymmetric 3D printed construct aimed for skin tissue regeneration. Colloids Surf, B 181:994–1003

112. Kim BS, Lee J-S, Gao G, Cho D-W (2017) Direct 3D cell-printing of human skin with functional transwell system. Biofabrication 9(2):025034

113. Koch L, Deiwick A, Schlie S, Michael S, Gruene M, Coger V, Zychlinski D, Schambach A, Reimers K, Vogt PM (2012) Skin tissue generation by laser cell printing. Biotechnol Bioeng 109(7):1855–1863

114. Frueh FS, Menger MD, Lindenblatt N, Giovanoli P, Laschke MW (2016) Current and emerging vascularization strategies in skin tissue engineering. Crit Rev Biotechnol

115. Risueño I, Valencia L, Jorcano J, Velasco D (2021) Skin-on-a-chip models: general overview and future perspectives. APL bioengineering 5(3):030901

116. Shahin H, Elmasry M, Steinvall I, Söberg F, El-Serafi A (2020) Vascularization is the next challenge for skin tissue engineering as a solution for burn management. Burns 8

117. Bertassoni LE, Cecconi M, Manoharan V, Nikkhah M, Hjortnaes J, Cristino AL, Barabaschi G, Demarchi D, Dokmeci MR, Yang Y (2014) Hydrogel bioprinted microchannel networks for vascularization XE "vascularization" of tissue engineering constructs. Lab Chip 14(13):2202–2211

118. Mori N, Morimoto Y, Takeuchi S (2017) Skin integrated with perfusable vascular channels on a chip. Biomaterials 116:48–56

119. Mori N, Morimoto Y, Takeuchi S (2018) Perfusable and stretchable 3D culture system for skin-equivalent. Biofabrication 11(1):011001

120. Tremblay P-L, Berthod F, Germain L, Auger FA (2005) In vitro evaluation of the angiostatic potential of drugs using an endothelialized tissue-engineered connective tissue. J Pharmacol Exp Ther 315(2):510–516

121. Marino D, Luginbühl J, Scola S, Meuli M, Reichmann E (2014) Bioengineering dermo-epidermal skin grafts with blood and lymphatic capillaries. Sci Transl Med 6(221) 221ra14–221ra14

122. Laschke MW, Vollmar B, Menger MD (2009) Inosculation: connecting the life-sustaining pipelines. Tissue Eng Part B Rev 15(4):455–465
123. Michael S, Sorg H, Peck C-T, Koch L, Deiwick A, Chichkov B, Vogt PM, Reimers K (2013) Tissue engineered skin substitutes created by laser-assisted bioprinting form skin-like structures in the dorsal skin fold chamber in mice. PLoS ONE 8(3):e57741
124. Liu X, Michael S, Bharti K, Ferrer M, Song MJ (2020) A biofabricated vascularized skin model of atopic dermatitis for preclinical studies. Biofabrication 12(3):035002
125. Keirouz A, Chung M, Kwon J, Fortunato G, Radacsi N (2020) 2D and 3D electrospinning XE "electrospinning" technologies for the fabrication of nanofibrous scaffolds for skin tissue engineering: a review. Wiley Interdisc Rev: Nanomed Nanobiotechnol 12(4):e1626
126. Keirouz A, Zakharova M, Kwon J, Robert C, Koutsos V, Callanan A, Chen X, Fortunato G, Radacsi N (2020) High-throughput production of silk fibroin-based electrospun fibers as biomaterial for skin tissue engineering applications. Mater Sci Eng, C 112:110939
127. Huang C-Y, Hu K-H, Wei Z-H (2016) Comparison of cell behavior on pva/pva-gelatin XE "gelatin" electrospun nanofibers with random and aligned configuration. Sci Rep 6(1):1–8
128. Dhandayuthapani B, Yoshida Y, Maekawa T, Kumar DS (2011) Polymeric scaffolds in tissue engineering application: a review. Int J Polym Sci
129. Babaeijandaghi F, Shabani I, Seyedjafari E, Naraghi ZS, Vasei M, Haddadi-Asl V, Hesari KK, Soleimani M (2010) Accelerated epidermal regeneration and improved dermal reconstruction achieved by polyethersulfone nanofibers. Tissue Eng Part A 16(11):3527–3536
130. Norouzi M, Soleimani M, Shabani I, Atyabi F, Ahvaz HH, Rashidi A (2013) Protein encapsulated in electrospun nanofibrous scaffolds for tissue engineering applications. Polym Int 62(8):1250–1256
131. Duan H, Feng B, Guo X, Wang J, Zhao L, Zhou G, Liu W, Cao Y, Zhang WJ (2013) Engineering of epidermis XE "epidermis" skin grafts XE "skin grafts" using electrospun nanofibrous gelatin XE "gelatin"/polycaprolactone membranes. Int J Nanomed 8:2077
132. Joseph B, Augustine R, Kalarikkal N, Thomas S, Seantier B, Grohens Y (2019) Recent advances in electrospun polycaprolactone based scaffolds for wound healing XE "wound healing" and skin bioengineering applications. Mater Today Commun 19:319–335
133. Milan PB, Lotfibakhshaiesh N, Joghataie M, Ai J, Pazouki A, Kaplan D, Kargozar S, Amini N, Hamblin M, Mozafari M (2016) Accelerated wound healing XE "wound healing" in a diabetic rat model using decellularized dermal matrix and human umbilical cord perivascular cells. Acta Biomater 45:234–246
134. Haldar S, Sharma A, Gupta S, Chauhan S, Roy P, Lahiri D (2019) Bioengineered smart trilayer skin tissue substitute for efficient deep wound healing XE "wound healing." Mater Sci Eng: C 105:110140
135. Savoji H, Maire M, Lequoy P, Liberelle B, De Crescenzo G, Ajji A, Wertheimer MR, Lerouge S (2017) Combining electrospun fiber mats and bioactive coatings for vascular graft prostheses. Biomacromol 18(1):303–310
136. Townsend-Nicholson A, Jayasinghe SN (2006) Cell electrospinning XE "electrospinning": a unique biotechnique for encapsulating living organisms for generating active biological microthreads/scaffolds. Biomacromol 7(12):3364–3369
137. Sampson SL, Saraiva L, Gustafsson K, Jayasinghe SN, Robertson BD (2014) Cell electrospinning XE "electrospinning": an in vitro and in vivo study. Small 10(1):78–82
138. Leong WS, Wu SC, Ng K, Tan LP (2016) Electrospun 3D multi-scale fibrous scaffold for enhanced human dermal fibroblast infiltration. Int J Bioprinting 2(1)
139. Pfuhler S, Fellows M, van Benthem J, Corvi R, Curren R, Dearfield K, Fowler P, Frötschl R, Elhajouji A, Le Hégarat L (2011) In vitro genotoxicity XE "genotoxicity" test approaches with better predictivity: summary of an IWGT workshop. Mutat Res/Genet Toxicol Environ Mutagenesis 723(2):101–107
140. Jusoh N, Ko J, Jeon NL (2019) Microfluidics-based skin irritation XE "irritation" test using in vitro 3D angiogenesis XE "angiogenesis" platform. APL Bioengineering 3(3):036101

141. Wagner I, Materne E-M, Brincker S, Süßbier U, Frädrich C, Busek M, Sonntag F, Sakharov DA, Trushkin EV, Tonevitsky AG (2013) A dynamic multi-organ-chip for long-term cultivation and substance testing proven by 3D human liver and skin tissue co-culture. Lab Chip 13(18):3538–3547

142. Alexander FA, Eggert S, Wiest J (2018) Skin-on-a-chip: transepithelial electrical resistance and extracellular acidification measurements through an automated air-liquid interface XE "air-liquid interface." Genes 9(2):114

143. Ataç B, Wagner I, Horland R, Lauster R, Marx U, Tonevitsky AG, Azar RP, Lindner G (2013) Skin and hair on-a-chip: in vitro skin models versus ex vivo tissue maintenance with dynamic perfusion. Lab Chip 13(18):3555–3561

144. Bhatia SN, Ingber DE (2014) Microfluidic organs-on-chips. Nat Biotechnol 32(8):760–772

145. Lee S, Jin S-P, Kim YK, Sung GY, Chung JH, Sung JH (2017) Construction of 3D multicellular microfluidic chip for an in vitro skin model. Biomed Microdevice 19(2):22

146. Song HJ, Lim HY, Chun W, Choi KC, Sung JH, Sung GY (2017) Fabrication of a pumpless, microfluidic skin chip from different collagen XE "collagen" sources. J Ind Eng Chem 56:375–381

147. Song HJ, Lim HY, Chun W, Choi KC, Lee T-Y, Sung JH, Sung GY (2018) Development of 3D skin-equivalent in a pump-less microfluidic chip. J Ind Eng Chem 60:355–359

148. Jeon HM, Kim K, Choi KC, Sung GY (2020) Side-effect test of sorafenib using 3-D skin equivalent based on microfluidic skin-on-a-chip. J Ind Eng Chem 82:71–80

149. Lim HY, Kim J, Song HJ, Kim K, Choi KC, Park S, Sung GY (2018) Development of wrinkled skin-on-a-chip (WSOC) by cyclic uniaxial stretching. J Ind Eng Chem 68:238–245

150. Sriram G, Alberti M, Dancik Y, Wu B, Wu R, Feng Z, Ramasamy S, Bigliardi PL, Bigliardi-Qi M, Wang Z (2018) Full-thickness human skin-on-chip with enhanced epidermal morphogenesis and barrier XE "barrier" function. Mater Today Commun 21(4):326–340

151. Ramadan Q, Ting FCW (2016) In vitro micro-physiological immune-competent model of the human skin. Lab Chip 16(10):1899–1908

152. Wufuer M, Lee G, Hur W, Jeon B, Kim BJ, Choi TH, Lee S (2016) Skin-on-a-chip model simulating inflammation, edema and drug-based treatment. Sci Rep 6(1):1–12

153. Agarwal T, Narayana GH, Banerjee I (2019) Keratinocytes XE "keratinocytes" are mechanoresponsive to the microflow-induced shear stress. Cytoskeleton 76(2):209–218

154. Kippenberger S, Bernd A, Guschel M, Müller J, Kaufmann R, Loitsch S, Bereiter-Hahn J (2000) Signaling of mechanical stretch in human keratinocytes XE "keratinocytes" via MAP kinases. J Investig Dermatol 114(3):408–412

155. Gupta P, SN GHN, Kasiviswanathan U, Agarwal T, Senthilguru K, Mukhopadhyay D, Pal K, Giri S, Maiti TK, Banerjee I (2016) Substrate stiffness does affect the fate of human keratinocytes, RSC Advances 6(5):3539–3551

156. Kenny FN, Drymoussi Z, Delaine-Smith R, Kao AP, Laly AC, Knight MM, Philpott MP, Connelly JT (2018) Tissue stiffening promotes keratinocyte proliferation through activation of epidermal growth factor signaling. J Cell Sci 131(10):jcs215780

157. Takei T, Han O, Ikeda M, Male P, Mills I, Sumpio BE (1997) Cyclic strain stimulates isoform-specific PKC activation and translocation in cultured human keratinocytes XE "keratinocytes." J Cell Biochem 67(3):327–337

158. Parmaksiz M, Dogan A, Odabas S, Elçin AE, Elçin YM (2016) Clinical applications of decellularized extracellular matrices for tissue engineering and regenerative medicine. Biomed Mater 11(2):022003

159. Badylak SF (2014) Decellularized allogeneic and xenogeneic tissue as a bioscaffold for regenerative medicine: factors that influence the host response. Ann Biomed Eng 42(7):1517–1527

160. Xie R, Zheng W, Guan L, Ai Y, Liang Q (2020) Engineering of hydrogel XE "hydrogel" materials with perfusable microchannels for building vascularized tissues. Small 16(15):1902838

161. Matei A-E, Chen C-W, Kiesewetter L, Györfi, A-H, Li Y-N, Trinh-Minh T, Xu X, Manh CT, Kuppevelt van T, Hansmann J (2019) Vascularised human skin equivalents as a novel in vitro model of skin fibrosis and platform for testing of antifibrotic drugs. Ann Rheum Dis 78(12):1686–1692

Chapter 4
Growing Skin-Like Tissue

Skin substitutes offer new therapeutic options for treating acute and chronic skin wounds [1, 2]. A functional skin substitute should replicate the morphological and biomechanical features and be able to mimic the gradients of various GFs, cytokines, enzymes, and pharmacological agents in vivo to promote optimal repair and regeneration of full-thickness skin defects [1, 3]. To achieve this, scientists have used natural and synthetic polymers to mimic the native ECM and recapitulate the structure and function of the targeted tissue [1, 4]. To even partially recreate a complex skin structure in vitro, three main elements must be considered:

(i)   cells,
(ii) matrices as support material for the placement of cells on or in them, and
(iii) GFs for tissue development [5–9].

The second point, the use of living cells seeded onto a natural or synthetic extracellular substrate to create implantable parts of organs, is more frequently associated with the concept of TE [7, 10]. This approach requires the development of model systems that better mimic the 3D (patho)physiological tissue microenvironment, including stiffness, topography, and biochemistry, and provide the ability to perform prolonged culture experiments while maintaining the desired tissue function [11]. As a complex material with highly variable properties in the different layers, the behaviour of skin is viscoelastic and anisotropic and can be influenced by the underlying tissue and bone. Variation in the mechanical properties of skin is also influenced by anatomical location and age, gender, level of care, and hydration. Therefore, different values of elastic modulus for different skin layers are reported in the literature [12, 13], and are listed in Table 4.1.
Another key factor to considered is the microstructure of the skin construct. Porous networks allow cells to attach to and penetrate the heterogeneous matrix while leaving room for further cell proliferation, efficient nutrient exchange, and good ventilation. Therefore, pore size and scaffold structure are essential for host tissue formation [6, 14–18]. Biodegradability may allow the scaffold to make room for the regenerated skin, but this property is not essential if a functional bioengineered skin is to

T. Zidarič et al., *Function-Oriented Bioengineered Skin Equivalents*,
Biobased Polymers, https://doi.org/10.1007/978-3-031-21298-7_4

Table 4.1 Elastic modulus of the different layers in human skin (based on [12, 13])

Skin layer		Elastic modulus
Stratum corneum	Dry	500 (3.5–1000) MPa
	Hydrated	30 (10–50) MPa
Viable epidermis		1.5 MPa
Dermis		20 (8–35) kPa
Hypodermis		2 kPa

be achieved. Regardless, they should maintain their 3D structure for at least three weeks to allow ingrowths of blood vessels, fibroblasts, and epithelial cell proliferation [6, 14, 19].

4.1 Materials

Selecting a suitable biomaterial as the support for tissue growth is the first step. State of the art in biomaterial design has considerably advanced over the past few decades. Biomaterials for biomedical applications aim to renew or restore the function of diseased or traumatized tissues in the human body [7, 14]. For skin tissue engineering, selecting the appropriate biomaterial should be based on the following criteria: degradability, mechanical strength, and biocompatibility with the immobilized cells. In addition, it should promote cell attachment and migration and be suitable for incorporating other materials and bioactive agents that provide functional and/or structural support to the reconstructed tissue [6, 14]. As mentioned above, the material's microstructure is one of the criteria for the construction of a skin equivalent. A highly porous structure (>90%) with well interconnected open pores (pore size 100–200 μm) is required to enable high cell density and efficient nutrient and oxygen supply to the cells [15]. As it seems, such interconnected 3D structure could be crucial to successfully grow thick tissue sections by providing immediate oxygen and nutrient supply [20]. Therefore, the materials used for engineered skin constructs tend to have inherent porous networks, such as sponges and hydrogels [14, 21]. Sponges are soft and flexible 3D polymer scaffolds with interconnected porous structures. Their porous network simulates the architecture of the ECM and allows cells to interact effectively with their environment. Due to their good fluid absorption capacity, they are considered suitable candidates for wound management. In addition, collagen-based sponges are widely used as insoluble 3D scaffolds for the fabrication of artificial dermis or skin [22–25]. Although sponges are mechanically more stable, their use is still limited due to open spaces distributed throughout the scaffold [7, 26]. Hydrogels are polymeric 3D networks with a hydrophilic nature that are cross-linked to form sponge-like matrices with high water content. Due to their remarkable properties, including a highly hydrophilic nature, similarity to native ECM, tunable physicochemical properties, biocompatibility, and biodegradability, they have facilitated progress in controlled drug delivery and TE applications

[5, 7, 27]. Both naturally derived (gelatin, collagen, fibrin, HA, CS, ALG, etc.) and synthetic (poly(ethylene glycol), PEG, poly(acrylamide), PAM, poly(2-hydroxyethyl methacrylate), PHEMA, etc.) hydrogels have been used for soft TE (summarized in Table 4.3). Still, natural hydrogels have potential inherent advantages such as biocompatibility, cell-controlled degradability, and intrinsic cellular interaction [5, 7, 28]. However, they may exhibit batch variations and normally have a narrow and limited range of mechanical properties. On the other hand, synthetic hydrogels offer the ability to precisely control structures and functions by means of chemical design of the monomer and polymer composition. Regardless of their origin, hydrogels are often favourable for promoting cell migration, angiogenesis, high water content, and rapid nutrient diffusion than synthetic polymers [7, 29]. In addition, hydrogels functionalized with GFs can directly support the development and differentiation of cells in the newly formed tissues [7, 30]. The commoly used polymers in skin tissue engineering are summerized in Table 4.2.

4.1.1 Collagen

Collagen is one of the most abundant proteins in human tissues and consists of three polypeptide chains with a helical structure (Fig. 4.1) [5, 33, 34]. There are many different types of collagens depending on their 3D structure and amino acid sequence (commonly repeated sequence of glycine, proline and hydroxyproline) that determine the functional requirements of various tissues [8, 33–35].

The long collagen fibrils can form bundles with much larger diameters, the thickness of which determines the tensile strength and resistance to enzymatic degradation of the connective tissue of which they are part. The orientation of the fibrils in the ECM is controlled locally by the cells that secrete them and influences cell migration and proliferation [34, 36, 37]. Because collagen is the predominant protein in natural skin (it makes up about 70% of body protein), it has been used extensively for skin engineering applications [9, 33, 38, 39]. It provides structural and biological support for cells through its fibril structure, which contains arginine-glycine-aspartic acid (RGD) residues that are known to boost cell adhesion [5, 40, 41]. In the 1980s, Burke et al. created a collagen-based artificial skin to treat extensive burn injuries [42], which was eventually commercialized as the Integra® Dermal Regeneration Template [5]. At the same time, Bell and colleagues began the development of a human skin equivalent, which was crucial to the development of Apligraf® [43]. Human fibroblasts taken from the neonatal foreskin were cast into a bovine collagen lattice seeded with epidermal cells to form a bilayer skin equivalent. This organotypic skin model was the first of its kind to contain both epidermal and dermal components, each of which housed living cells [43, 44]. In the late 1980s, another group of researchers also fabricated a collagen-GAG scaffold that contained autologous fibroblasts with a stratified layer of cultured autologous keratinocytes [45]. In subsequent years, clinicians have used collagen scaffolds extensively in wound healing

Table 4.2 Natural, synthetic and composite[1] (blend) polymers commonly used for skin TE applications and their specific characteristics (adapted based on [2, 31])

Polymer	Type	Specific features
Collagen	Natural	Reduced wound contraction compared to freeze-dried scaffold materials Higher cell proliferation in aligned fibres compared to the random scaffold
Collagen/Silk fibrin	Natural blend	Better cellular responses in hybrid nanofibrous scaffolds compare to blend nanofibrous scaffolds
Gelatin	Natural	Well-stratified dermal and epidermal layers including a continuous basal keratinocyte layer Promotes cell adhesion and spreading of type I collagen, suitable for wound dressing
Gelatin/chitosan	Natural blend[2]	High water uptake capacity Positive influence on proliferation and adhesion of keratinocyte and fibroblast co-culture Reduced scar formation
Chitosan	Natural	Activates platelets and accelerates natural blood clogging Accelerates wound repair by stimulating fibroblast proliferation, macrophage activation, cytokine production, angiogenesis and promotes collagen deposition
Chitosan/ Collagen	Natural blend[2]	Keratinocyte migration and wound re-epithelization, prerequisite for healing and regeneration
Fibrin	Natural	Provides a mechanically ECM, suitable for long-term growth and differentiation of keratinocytes Ability to prepare of a large surface of stratified epithelium on the dermal equivalent
Hyaluronic acid	Natural	High water uptake capacity Associated faster wound healing and reduced scar formation
PCL[4]	Synthetic	Enhanced cell proliferation and wound healing (formation of hydroxyl and carboxyl groups that promotes cell proliferation)
PLLA[5]	Synthetic	Enhanced epidermal skin cell migration across the wound
PLGA[6]	Synthetic	Highly porous scaffold provides mechanical support for cells to maintain uniform distribution Improved cell migration, and collagen secretion
PCL/Chitosan	Mixed blend[3]	The presence of chitosan improved the hydrophilicity of the scaffold, bioactivity and protein adsorption
PCL/Collagen	Mixed blend[3]	Supports the attachment and proliferation of human dermal fibroblasts Suitable for dermal substitutes
PCL/Gelatin	Mixed blend[3]	Increased cellular adhesion and metabolism
PLGA/Chitosan	Mixed blend[3]	Improved cytocompatibility compared to pure PLGA

(continued)

Table 4.2 (continued)

Polymer	Type	Specific features
PLGA/Collagen	Mixed blend[3]	Enhanced cell attachment, proliferation and ECM secretion Effective as wound-healing accelerators in early-stage wound healing

[1]Composite—materials in which the individual phases are separated at the molecular level and in which properties such as the modulus of elasticity change significantly compared to those of a homogeneous material [32]; [2]Natural blend—combination of different natural polymers; [3]Mixed blend—combination of natural and synthetic polymer; [4]PCL—poly(ε-caprolactone); [5]PLLA—poly(L-lactic acid); [6]PLGA—poly(lactic co-glycolic acid)

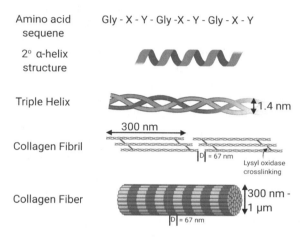

Fig. 4.1 Structure of collagen. The amino acid sequence of collagen consists of a repeating sequence Gly-X–Y, where X and Y are usually proline and hydroxyproline. This unique sequence is responsible for the α-helical secondary structure of collagen. Fibrillar collagen is a triple helix containing cross-links formed by the action of lysyl oxidase. In vivo, these collagen fibrils form fibers with varying thickness and a periodic repeat (D) of 67 nm. Reproduced from [34] with permission from MDPI

under various commercial brands [5, 8, 46, 47]. Collagenous products are valuable due to their general biocompatibility and desired biodegradability. Therefore, collagen is inevitable for many TE applications, especially for wound healing and skin TE [8, 33]. For example, Chan et al. used bovine type I collagen to construct a 3D scaffold with uniform pore size to investigate its ability to support in vitro vascularization with human endothelial cells and facilitate extrinsic/intrinsic vascularization in vivo. Using two different animal models, the mouse hypodermis implant model for extrinsic vascularization and the rat chamber model for intrinsic vascularization. The authors successfully demonstrated that collagen is suitable for both types of vascularization [48]. Although collagen-based materials still occupy the dominant position in the skin TE, their rapid degradation and consequent unstable mechanical structure have led to undesirable outcomes [5, 36]. To reduce the rate of degradation and improve the mechanical properties and applicability of collagen scaffolds fabricated by either conventional (casting spongy collagen scaffolds) or rapid biofabrication methods (3D bioprinting, electrospinning), researchers have employed biocompatible crosslinking strategies and/or blended them with other biomaterials, generally more robust synthetic polymers [5, 8, 34, 38, 39, 49–56]. Most chemical cross-linking in collagen is achieved by the specific use of amine or carboxylic acid groups in amino acid side chains. To improve and stabilize the mechanical and biological properties of collagen, the following crosslinking agents are commonly used: 1-ethyl-3-(3-dimethylaminopropyl)carbodiimide/N-hydroxysuccinimide (EDC/NHS), glutaraldehyde, and genipin [34, 57]. In the case of genipin, crosslinking also alters the inherent fluorescence of the collagen and largely eliminates the characteristic fibrillar streaks of the collagen [34, 58]. Glutaraldehyde has also been used as a crosslinking agent; however, the cytotoxicity of glutaraldehyde residues and overcrosslinking limit the application of modern collagen-based products [34]. In addition to chemical crosslinking, physical crosslinking methods such as dehydrothermal (DHT) treatment and UV irradiation have also been used to crosslink collagen [34, 57]. Although all crosslinking methods increase the tensile strength of hydrogels and decrease the enzymatic degradation rate, previous studies suggest that physical crosslinking (DHT treatment and UV irradiation) impairs cell migration [34, 59].

Instead of cross-linking, blending collagen with synthetic polymers (e.g., PCL, PLA, PEG, PLGA, etc.) enables the improvement of optimal mechanical and biological properties of scaffolds for specific engineering applications. In this context, the synthetic polymer provides mechanical support for the structure of the scaffolds, while the collagen provides cell recognition signals on the surface and inside the scaffolds, which are crucial for cell behaviour and development [60, 61].

4.1.2 Gelatin

Gelatin is the irreversibly denatured form of collagen obtained by partial hydrolysis [5]. It contains high levels of glycine, proline and hydroxyproline, and as with collagen, the presence of RGD residues promotes cell adhesion, proliferation and migration [5, 62]. In contrast to collagen, gelatin is fully absorbable in vivo. It further exhibits lower antigenicity and has a high hemostatic effect [2, 5, 8, 22, 63]. Collagen and its derivatives, including gelatin, can attract fibroblasts in vivo during wound and fracture healing, as well as embryogenesis [22]. Since gelatin has been shown to have a similar chemical structure to GAG and collagen, some efforts have been made to use gelatin as a supportive scaffold to mimic the native ECM in skin TE [46, 64]. As a component of artificial skin, Lee and co-workers prepared a gelatin/β-glucan sponge containing fibroblasts, using gelatin to promote epithelialization and granulation tissue formation and β-glucan to boost the immune system by activating macrophages. To develop a stratified wound dressing that mimics normal skin, they cultured a multilayered structure of fibroblasts and keratinocytes at the air–liquid interface in vitro. The collagen synthesized by the fibroblasts improved the attachment of keratinocytes to the surface of the artificial dermis. Moreover, the fibroblasts on the dermis improved the re-epithelialization in vivo on the full-thickness skin defect [22]. Under certain conditions, such as temperature, pH, or exposure to organic solvents, gelatin macromolecules exhibit sufficient flexibility to realize different conformations; so, the gelatin characteristics depend on their molecular structure. Differences in collagen sources and preparation methods result in a structure with variable physical properties and chemical heterogeneity. For example, both acidic and basic functional groups in gelatin macromolecules contribute to greater structural diversity compared to other natural or synthetic polymers [8, 9, 65, 66]. Although it is a popular biomaterial for many biomedical applications (drug delivery systems, TE), it suffers from poor mechanical properties such as low tensile strength and thermal stability and faster degradation rates [5, 67–70]. Gelatin hydrogels lack thermal stability and must be chemically crosslinked if they are intended to be used as tissue culture materials. Almost all crosslinking methods for collagen also apply to gelatin [36, 71–73]. Covalent crosslinking increases the strength of gelatin hydrogels and inhibits enzymatic degradation, which can be adjusted by controlling the degree of crosslinking. Alternatively, gelatin can be photo-crosslinked with methacrylate or free thiol groups after functionalization. Both 2D and 3D cell cultures of endothelial cells in photo-crosslinked gelatin methacrylate (GelMA) showed good adhesion, proliferation, elongation, and migration. This modified gelatin possesses RGD sequences and matrix metalloproteinase (MMP) sequences that support enzymatic degradation and play critical roles in dermal wound healing, morphogenesis, and tissue repair [70, 74, 75]. Still, in 3D cultures, elongation of cells and formation of interconnective networks may be hindered by the stiffer gel [36, 65, 76]. Zhao and colleagues synthesized photo-crosslinked GelMA hydrogels with tunable mechanical and enzymatic degradation properties, which were ideal as skin scaffolds TE. By varying the concentration of the GelMA prepolymer solution, they were able to tune the physical and biological features to meet the needs of epidermis formation [73]. An epidermis equivalent containing immortalized human keratinocytes (HaCaT) in GelMA at higher concentrations exhibited enhanced material stiffness with prolonged resistance to enzymatic degradation compared to collagen-based substitutes. They were shown to support the development of multilayered, renewable tissue constructs with similar organization and differentiation as found in the human dermis. In addition, the increased stiffness also contributed to the formation of a stratified epidermis with some barrier function, including electrical resistance and prevention of water loss [5, 73]. Polymer blending is a widely used technique for developing novel materials. Polymer blends refer to a polymeric material composed of at least two polymers, resulting in improved physicochemical properties compared to those of a single polymer. Broadly speaking, they can be classified as either miscible (homogeneous) or immiscible (heterogeneous) blends, depending on the interaction behavior of the polymers that make up the blend. The thermodynamics of polymer systems play an important role in determining the miscibility of polymer blends. Miscible blends have similar properties comparable to random copolymers or homopolymers. In contrast, immiscible blends have multiple glass transition temperatures due to the distinct separation between the individual polymers. The preparation of blends is usually based on physical blending such as extrusion, mixing, and injection molding, and a variety of properties can be achieved by a suitable combination of pure components. In addition, the pore size and porosity of the scaffold can be easily controlled by adjusting the parameters of the blending technique used. However, the main challenge in polymer blending is the thermodynamic heteogeneity of most polymers, which leads to phase separation in the absence of mechanical stimulus. To overcome this problem, a compatibilizer is usually used to reduce the interfacial tension and thus increase the interaction forces between the individual polymers. In fact, it is important that the blends remain stable to create functional materials with the desired properties [77–79]. Gelatin has numerous reactive sites (amine and carboxyl groups) that allow crosslinking, functionalization, and even attachment of hydrophobic materials to its backbone to improve physical and chemical stability, including water resilience and thermos mechanical performance, for biomedical applications [80]. This unique character when blended with other natural and synthetic polymers, contributes to the improved biological properties, in particular cell attachment due to the RGD sequences, of the fabricated scaffold [8, 66, 80]. In this context, Wang et al. [62] developed a bilayer gelatin-6-sulphate hyaluronic acid (gelatin-C6S- HA) to mimic the ECM composition of the skin. Here, gelatin was used as the main matrix for scaffold design to support cell attachment and nutrient delivery, while the incorporated HA and C6S provide a better microenvironment for cell proliferation, differentiation and migration. It has been reported that HA promotes cell migration, which resulted in better cell distribution in the porous scaffold in this study (Fig. 4.2).

In addition, both HA and C6S can suppress macrophage multinucleation, resulting in a reduced inflammatory response. The lower membrane layer with larger pores (150 μm) was seeded with fibroblasts for the dermal component and also served as a feeder layer for inoculation of keratinocytes. The latter were seeded and cultured on the upper layer with smaller pores (20–50 μm), first in submerged conditions and then at an air–liquid interface. This newly designed skin equivalent formed well within three weeks. It was characterized by an epidermis-like structure that included the suprabasal layers and the stratified layer and a dermis structure with sparse distribution of fibroblasts surrounded by their own ECM [62]. Preliminary studies have shown that the polymer blend of PCL/gelatin is an uncompromising solution to overcome the shortcomings of natural and synthetic polymers and creates a new biomaterial with good biocompatibility [81]. Based on these findings, Chong and colleagues [82] fabricated a nanofibrous scaffold from a polymer blend of PCL/gelatin. They presented a novel cell cultivation technique in which fibroblasts were seeded on both sides of an electrospun nanofibrous scaffold to create a 3D dermal analogue populated with fibroblasts. To achieve maximum cellular ingrowth, cells must migrate into the scaffold structure rather than just onto the substrate surface. As part of developing a 3D scaffold for fibroblast infiltration, the authors successfully cultured fibroblast cells on both sides of a thin PCL/gelatin nanofiber scaffold. This allowed the growth

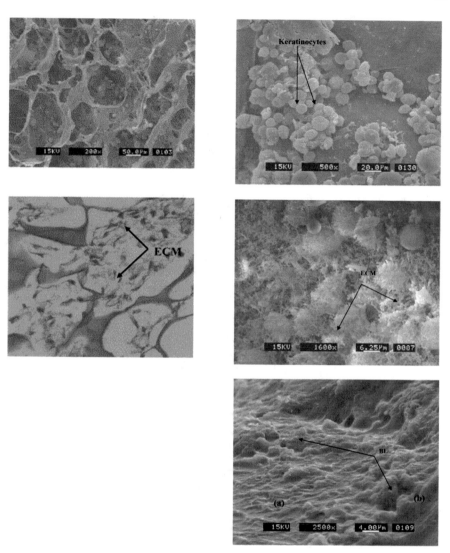

Fig. 4.2 SEM images of seeding, migration and proliferation of dermal fibroblasts (left) and keratinocytes (right). The obtained micrographs show that the pore size and porosity of the scaffold play an important role in the communication between keratinocytes and dermal fibroblasts, which help keratinocytes to differentiate into a multilayered epidermis. Dermal fibroblasts seeded on the thicker lower layer became trapped in the pores of this layer, proliferated, secreted ECM proteins, and later developed into the dermis. Adapted from [62] with permission from John Wiley & Sons, Inc

and proliferation of cells on both sides of the scaffold and maximized the cell loading into a stereostructure, resulting in close biomimicry of the natural ECM [82].

4.1.3   Fibrin

Fibrin is a naturally occurring protein in the blood characterized by excellent biocompatibility and biological properties. It is a hemostatic substance activated after injury and serves as structural support in wound healing [5, 83]. As a biomaterial, it has shown great potential in recent years as a tissue culture model to study cellular behaviour in 3D and as an injectable scaffold in TE [5, 36]. Fibrin is formed by thrombin-triggered aggregation of insoluble polypeptide chains of fibrinogen into a network of fibrils (Fig. 4.3) [36, 84].

Similar to collagen- and gelatin-based skin constructs, fibrin scaffolds contain natural cell-binding sites for cell adhesion, with the distinct advantage of not requiring further chemical modification to incorporate these sites into the hydrogel formulation [5, 9, 36, 85]. It binds many GFs and coagulation components such as fibronectin and

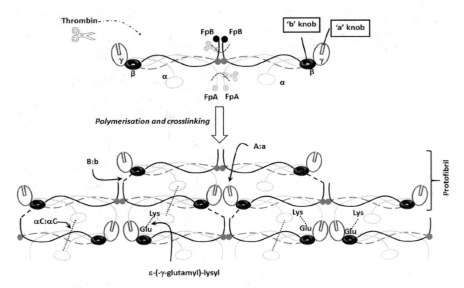

Fig. 4.3 Fibrin polymerization. Thrombin-mediated cleavage of fibrinopeptides FpA and FpB initiates the conversion of fibrinogen to fibrin. Cleavage of FpA and FpB exposes knob 'A' and 'B', which are complementary to pockets 'a' and 'b' located on the γ- and β-nodules of an adjacent fibrin monomer. During and after the polymerization process, fibrin is covalently cross-linked by thrombin-activated factor XIIIa (FXIIIa), which catalyses the formation of ε-(-γ-glutamyl)-lysyl cross-links between lysine and glutamate residues in the γ-chain, thereby increasing fibre density and stiffness. Reproduced from [84] with permission from MDPI

HA. It has two pairs of RGD sequences and one pair of alanine-glycine-aspartic acid-valine (AGDV) sequences that can react with cell surface integrins [36]. The fibrin network exhibits unique viscoelastic properties that differ in many aspects from those of synthetic hydrogels. Fibrin-based hydrogels exhibit nonlinear elasticity; the elastic modulus of fibrin increases sharply the more the material is deformed [36, 86]. Fibrin was used as an injectable biomaterial for in situ formation of 3D ECM containing various types of cells, including keratinocytes [87, 88], fibroblasts [89, 90], and MSCs [91]. Based on previous reports on the use of fibrin gels to promote fibroblast and endothelial cell growth, Meana et al. [88] developed a new keratinocyte culture system on a fibrin-based dermal equivalent for skin wound closure. The dermal matrix of fibrin gel contained cultured human fibroblasts from the foreskin. In different seeding ratios, a secondary culture of keratinocytes was seeded on the fibrin-fibroblast gel. The biochemical and cellular composition of the described fibrin-fibroblast gel, which was very similar to that of the skin during the initial steps of wound healing, allowed the production of a large surface area of stratified epithelium on the dermal equivalent [88]. Gorodetsky and his team [89] developed fibrin-derived microbeads (FMB), 50–200 μm in diameter, that served as effective cell vehicles and matrices for cell growth. Polymeric microbeads are popular as cell or drug carriers due to their ease of preparation and administration. The term microbead describes an aspherical particle with a diameter between 0.5 and 1000 μm [92]. In the above study, the porosity of FMB provides a large surface area for cell attachment and stimulates cell migration and proliferation due to reduced contact inhibition. In the porcine skin wound model, FMB seeded with fibroblasts significantly accelerated the healing process and granulation tissue formation. The benefit of fibroblasts transplanted on FMB may be due in part to their dense composition, which facilitates attachment to the wound bed and allows proper cell alignment [89]. Compared to collagen, fibrin provides a mechanically stable and rigid scaffold that generates intrinsic tension that stimulates fibroblasts to actively synthesize and organize their ECM. This newly synthesized authentic ECM is in turn suitable for long-term keratinocyte growth and differentiation. In addition to its initial function of mechanically stabilizing fibroblasts in dermal equivalents, it shows a marked positive effect on seeded keratinocytes and subsequent epidermal morphogenesis. The functional role of fibrin as a scaffold biomaterial and its influence on ECM production is still under speculation. It is likely that a major effect of the fibrin scaffold is that it provides a sufficiently stable and rigid framework, resulting in an intrinsic tension within dermal equivalent that stimulates fibroblasts to actively synthesize and organize their authentic ECM. The in vivo-like architecture sets the stage for the formation of a mature epidermal basement membrane. The correct assembly and spatial organization of the basement membrane leads to a regular integrin pattern in basal keratinocytes, which is involved in the control of epidermal differentiation, proliferation and morphogenesis [93].

4.1.4 Hyaluronic Acid

Hyaluronic acid (hyaluronan, HA), an immunoneutral polysaccharide, is the only nonsulfated GAG widely distributed in the body [36, 94, 95]. The identical structure of the molecule in all living organisms, associated with the minimal risk of immunogenicity, makes it an almost ideal clinical biomaterial [96]. As a member of the group of GAGs, it is an essential component of the ECM. It is one of the key components in wound healing, cell motility, angiogenesis, cellular signalling, and matrix organization [26, 36, 62, 94, 95]. It has the great advantage of being able to indirectly stimulate neocollagenogenesis (i.e., new production of collagen) through the mechanical stretching of the dermis and subsequent activation of dermal fibroblasts [96, 97]. Its high molecular weight (100 kDa in serum to 8000 MDa in dentin [98]), associated unique viscoelastic and rheological properties make HA an attractive scaffold material for whole skin or just dermis engineering [94, 95]. Under the action of a shear force, the viscosity of the polymer solution decreases markedly with increasing shear rate and is characterized by a non-Newtonian fluid [99]. In addition, HA has a high water-binding capacity, which provides a moist environment that protects the injured tissue surface from drying out and can modulate the cellular microenvironment [26, 95, 100]. Moreover, HA has shown beneficial effects in scarless wound healing. One of its degradation products (N-acetyl-D-glucosamine) promotes fibroblast proliferation and orderly collagen deposition, resulting in faster wound healing and less scarring [26, 101]. The molecular weight of HA is a key factor in the growth of skin cells or tissue. A higher molecular weight HA blocks endothelial cell migration and angiogenesis. On the other hand, HA with a lower molecular weight has pro-inflammatory (adjusting phagocytosis by regulating the movement of macrophages) and pro-angiogenic (enhancing the mitosis of endothelial cells) activities [2, 94]. Its known ability to heal wounds faster and without scarring makes HA a commonly used biomaterial to manufacture wound care products. Hyaff®, Laserskin® and Hyalograft® are currently commercial wound care products based on HA [2, 102, 103]. Several studies report the use of HA in various wound healing related applications. Anisha and colleagues [26] developed a nanocomposite CS/HA sponge as a potential wound dressing material. By combining the properties of CS and HA, the fabricated porous nanocomposite sponges exhibited controlled swelling and biodegradation accompanied by improved blood clotting and platelet activation. Moreover, these nanocomposite sponges increased the proliferation rate of human dermal fibroblasts, which exhibited elongated morphology after two days, pinpointing cell migration and, consequently, faster wound healing [26]. The beneficial effect on accelerated wound healing was also studied by Uppal et al. [104]. They compared the healing performance of different wound dressings, including an adhesive dressing, a solid HA, a gauze dressing coated with Vaseline, an antibiotic dressing, and a HA nanofibrous wound dressing. Preclinical evaluation on five pigs showed that epithelial tissue completely covered wounds treated with HA nanofibers. A HA nanofibrous dressing was also superior in wound healing rate, which was the fastest among all groups [104]. In addition, HA membranes have shown their merit as matrices for the growth and transfer of cultured keratinocytes to reconstruct full-thickness wounds [105]. Although skin replacement by cultured epithelial autograft (CEA) is a well-accepted method, several clinical drawbacks have led to various efforts to optimize cell culture and delivery. Among them is the transfer of human keratinocyte monolayers directly to the recipient's wound bed with the carrier material on top, known as "upside-down" grafting. This approach was described by Horch et al. [105] by using esterified HA membranes that served as carriers for the cultivation and transfer of human keratinocytes. Moreover, the unique physical (expansion during water uptake, improved resistance to enzymatic degradation) and biological properties (anti-adhesive and anti-inflammatory effects) of an esterified HA make it a promising biomaterial for enhancing tissue repair [105, 106]. According to the authors, the "upside-down" strategy guarantees optimal delivery of cultured cells to the wound site and optimal cell yield in the wound bed. The efficiency of the proposed grafting method using an esterified HA membrane was demonstrated by the successful transfer of proliferative keratinocytes with simultaneously reduced wound contraction. Furthermore, no relevant immunological reaction to HA based materials was reported, making these materials very interesting for additional advantages in establishing skin repair strategies [105].

4.1.5 Chitosan

Chitosan (CS) is a partially deacetylated derivative of chitin, occurring mainly in the exoskeletons of arthropods, fungi, and the crustaceans' shells. It is a linear polysaccharide consisting of $\beta(1-4)$-linked D-glucosamine residues with a variable number of randomly arranged N-acetyl-D-glucosamine groups (Fig. 4.4) [2, 5, 8, 9, 107, 108].

CS is popular in numerous applications as a scaffold for tissue culture and wound dressings due to its biodegradability and biocompatibility [8, 15, 107]. In addition, it is an analgesic and hemostatic polymer that can be modified to exhibit antimicrobial and anti-inflammatory properties [2, 5, 107, 109, 110]. CS-based scaffolds provide a suitable template and physical support to direct the differentiation and proliferation of cells into functional target tissues or organs. As a hemostatic agent,

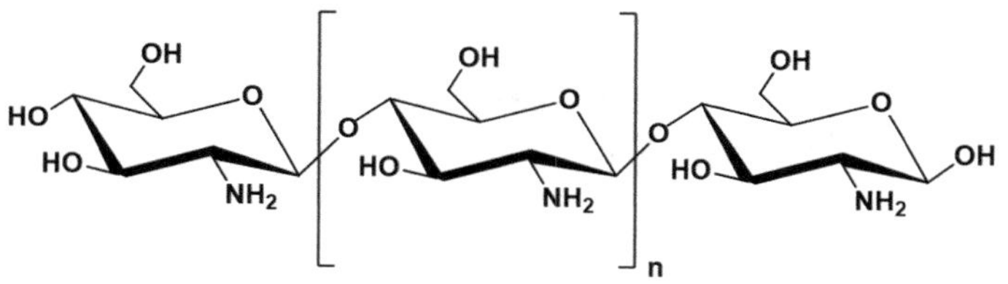

Fig. 4.4 Chemical structure of CS. Reproduced from [108] with permission from MDPI

it activates platelets and accelerates natural blood clogging [2, 8, 107]. It also accelerates wound healing by stimulating fibroblast proliferation, macrophage activation, cytokine production, and angiogenesis. It also promotes collagen deposition through the gradual depolymerization and release of N-acetyl-D-glucosamine (similar to HA) [2, 8, 26, 109]. In addition, it can also induce high levels of natural HA synthesis at the wound site, which in turn can improve wound healing and reduce scar formation [2, 109]. Ma and colleagues [111] created a bilayer CS material in the form of a film and a sponge (both CS-derived) to create a new artificial dermis and scaffold for the skin TE. The film layer was designed to control evaporation of body fluids similar to a standard wound dressing. In contrast, a positively charged sponge layer served as a scaffold for attachment and growth of human neofetal dermal fibroblasts. They prepared a dense CS film by casting and a porous CS sponge by freeze-drying. Dermal fibroblasts were seeded within the sponge layer of the bilayered CS material, which allowed cell growth and proliferation. Moreover, the fibroblasts could bind tightly with the sponge layer via the newly formed ECM, resulting in a living cell–matrix-CS composite. The resulting bilayer CS material maintained its stability in shape and size during cell culture [111]. Like other hydrogels, CS suffers from poor mechanical properties and slow gelation. Blending with other polymers or crosslinking are effective ways to obtain a suitable biomaterial for TE [5, 15]. To overcome the brittle behaviour of CS-based scaffold materials, Han et al. [15] combined CS and gelatin to prepare a gelatin/CS sponge scaffold (gelatin/CS) for skin repair. The gelatin/CS sponges were prepared by freeze-drying. They were designed as an absorption system with excellent water uptake capacity and a template to guide cell adhesion, expansion, proliferation and differentiation [15]. The positive influence on the co-culture of keratinocytes and fibroblasts in terms of proliferation and adhesion has also been demonstrated [5, 112]. In a rabbit model, CS sponges loaded with homologous fibroblasts and keratinocytes significantly reduced the production of collagen and α-smooth muscle actin (ASMA, a marker used to identify pathological fibroblasts) that drive fibrosis [113], thus reducing the tendency towards scar formation. However, it seems that the homologous fibroblast- and keratinocyte-loaded variants of the CS sponge had different healing abilities [112]. Several clinical studies have reported the favourable results of using CS for wound healing in patients who have undergone plastic surgery [114] and skin grafting [115, 116]. Currently, there are several CS-based wound dressings on the market in the form of nonwovens, nanofibers, composites, films, and sponges [109]. Tegasorb™ and Tegaderm™ wound dressings are used for the protective treatment of partial and full-thickness dermal ulcers, leg ulcers, superficial wounds, abrasions, burns as well as donor sites. The presented CS absorbs wound exudates, then swells and forms a soft gel mass that promotes wound healing [117].

4.1.6 Alginate

Alginic acid and its salts (usually the sodium salt) are polysaccharides widely used in the food industry and in the biomedical field [118–120]. ALG is a linear anionic polysaccharide containing homopolymeric blocks of 1,4-linked β-D-mannuronate (M) and α-L-guluronate (G), carrying free hydroxyl (-OH) and carboxyl (-COOH) functional groups (Fig. 4.5) [2, 5, 36, 120, 121].

These blocks are arranged either as homopolymeric (GG or MM) or heteropoly-meric (GM or MG) blocks. The relative amount of these monomers (G/M ratio) deter-mines the rigidity of the ALG polymer chain. A higher proportion of heteropolymeric units results in gels with lower tensile strength compared to ALG, which are rich in homopolymeric units. This is probably due to the sterically hindered packing of the adjacent chains and the subsequent formation of junction zones in the heteropoly-meric blocks [120, 121]. ALG is generally considered biocompatible, and gelation is induced by several mechanisms. Its intrinsic properties affect the gelation process differently. While the molecular weight of ALG can affect the viscosity and gelation kinetics of ALG solutions, the type, ratio and length of G/M units affect the gelation process. At acidic pH values (pH below 3), ALG self-assembles into acidic gels by forming intermolecular hydrogen bonds. In addition, at low pH, ALG, which is rich in MG /GM blocks, has higher solubility, while a higher content of both homopoly-meric (MM or GG) blocks contributes to the water insolubility of ALG. [36, 118, 120, 121]. Alternatively, the G units (but not exclusively) in the ALG allow the polymer to be cross-linked by divalent cations such as Ca^{2+} or Ba^{2+}, interacting with ALG free -COO⁻ groups. This ionic gelation process is known as the "egg-box" model (Fig. 4.6) [2, 36, 120, 121].

However, cation-crosslinked ALG gels have the disadvantage of uncontrolled degradation due to the diffusion/exchange of ions under physiological conditions. To address this problem, Leung et al. [122] proposed a dual crosslinking process using calcium and glutaraldehyde to improve the degradation resistance of ALG

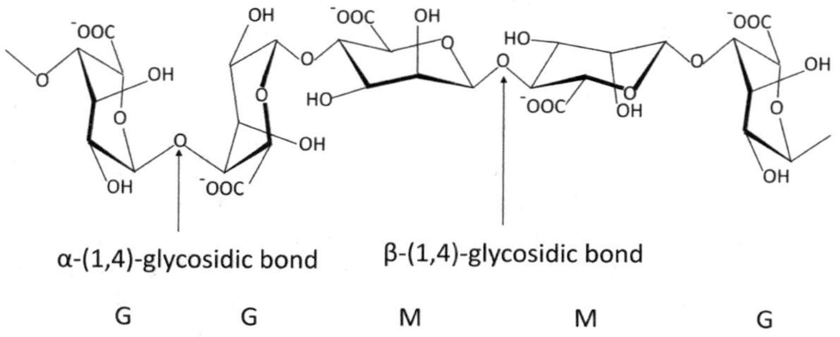

α-(1,4)-glycosidic bond   β-(1,4)-glycosidic bond

G   G   M   M   G

Fig. 4.5 ALG is composed of α-L-guluronic acid (G) and β-D-mannuronic acid (M) blocks. Different G and M blocks content and arrangement (homopolymer: GG, MM units or heteropolymer: GM or MG units) detrmines the ALG properties (rigidity of the polymer chain, affinity toward ions, viscosity, etc.). Adapted from [121] with permission from MDPI

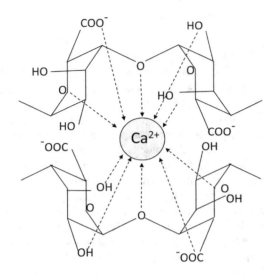

nanofibers in sodium-rich environments [2, 122]. Since ALG undergoes rapid but reversible ionic crosslinking, it has been used as a sacrificial template in TE. The main reasons for its popularity among tissue engineers are its low cost, rapid and straightforward gelation, and non-immunogenicity, which make it a suitable candidate for covering excisions after skin donor surgery [5, 123]. ALG forms a linear, water-rich polymer structure and is hydrophilic and viscous, making it particularly suitable for cell environments [94, 124–126]. Several studies have shown that ALG has a positive effect on wound healing due to its similarity to the native ECM of the skin in many respects [125–130]. Its hydrophilic nature is particularly well suited for the encapsulation of cells, GFs and even drugs for the regeneration of the skin or other tissues and/or engineering purposes [124, 125, 128–130]. Like HA, mammalian cells are less inclined to proliferate intensively when in contact with ALG-based materials (despite the proven biocompatibility of ALG with various cells [131]). This is due to the hydrophilic nature of alginic acid, which adsorbs fewer proteins, resulting in less adhesion and proliferation [36, 125]. Recently, our group introduced an ALG-based bioink with encapsulated human fibroblast (hSF) cells to fabricate a 3D bioprinted dermis-like structure [125]. To compensate the inability to maintain a uniform 3D structure of ALG scaffolds, the authors combined ALG with nanofibrillated cellulose (NFC) and carboxymethyl cellulose (CMC) in attempt to developed a novel bioink (ALG /CMC/NFC-bioink) by assimilating the shear-thinning properties of NFC hydrogels with the rapid crosslinking ability of ALG and CMC. The positive effect of using nanofibrillated cellulose (NFC) in soft TE applications can be attributed to its high hydration potential and its morphological similarity to ECM components such as collagen [132, 133]. The excellent printability of the developed hSF-loaded bioink enabled 3D bioprinting of complex scaffolds with controlled cell density and defined porosity. The 3D hSF-loaded scaffolds showed shape and size stability after the bioprinting process and after cultivation for up to 29 days. Their characterization also revealed high incorporated cell viability, cell proliferation, and a tendency to form agglomerates during cultivation in a constructed 3D environment [125]. Although the native ALG exhibits poor cell adhesion, the underlying hSFs with their secreted ECM proteins presumably alter the internal structure of the ALG/CMC/NFC scaffold, promoting the attachment of "new" cells [134, 135]. The anionic nature of ALG allows the formation of polyplexes with a cationic polymer, such as CS, which improves the mechanical properties of ALG hydrogels [36]. In this context, Kong and co-workers [136] fabricated a free-standing ultrathin CS/ALG membrane as a template for wound healing using the layer-by-layer (LBL) self-assembly technique. The theory behind this is that the amino group of CS enables it to ionize in water to form a polycation, which exerts a natural attraction to ALG as the carboxyl group of ALG simultaneously makes it a polyanion in water. As a result, this interaction creates the exact condition necessary to integrate the two polyelectrolytes on a given substrate through LBL self-assembly [8, 136]. Alternatively, the chemically modified ALG can significantly improve the mechanical stability of the 3D construct. The chemical methods commonly used (more detailed described by Rosiak et al. [137]) are: (i) acetylation, phosphorylation and sulfation of ALG -COOH groups, (ii) oxidized of ALG and reductive amination of oxidized alginate, (iii) esterification and amidation of ALG -COO$^-$ groups, (iv) ligation of bioactive ligands with ALG carboxyl groups, and (v) its modification by click chemistry reactions [121]. Its dual network composite with synthetic polymers, such as PEG, can create a tough and printable hydrogel that is even suitable for cell encapsulation and long-term survival within the bioprinted 3D architecture [138].

4.1.7 Decellularized Extracellular Matrix

Decellularized extracellular matrices (dECMs) are considered one of the most promising materials in TE as they preserve a complex of functional and structural proteins consisting of GAGs, glycoproteins and bioactive cues present in all the features of natural ECMs (Fig. 4.7) [139, 140].

In recent decades, dECMs have been used with considerable success in TE in various ways, from biological sheets to injectable hydrogels and coatings. Studies have demonstrated the enormous potential of dECM materials, indicating that natural bioactive molecules in a dECM could significantly restore tissue homeostasis and promote the regeneration process [140, 142]. Therefore, dECM materials could provide a suitable platform for cell behaviour control and regenerative therapies [139, 143]. Milan and colleagues [141] fabricated a cellular dermal substitute that provided a skin tissue-like specific environment for a cell delivery system. To promote angiogenesis and accelerate wound healing, they constructed a fibrous decellularized dermal matrix (DDM) that incorporated human umbilical cord perivascular cells (HUCPVCs), a type of MSCs that can stimulate early wound healing in skin defects. This type of scaffold could have a dual function: biomechanical support and

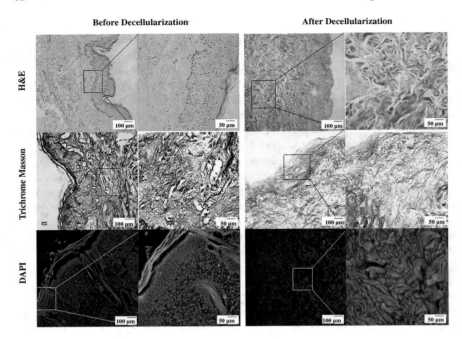

Fig. 4.7 Histological examination of human dermis before and after decellularization process. H&E and Masson's trichrome staining demonstrated that there are no cells or cellular debris and skin appendages (hair follicles, sweat glands, endocrine glands etc.) in decellularized dermal matrix (DDM) following the decellularization process. The 4',6-diamidino-2-phenylindole (DAPI) staining shows that the DDM scaffold is free of cellular nuclei while retaining structural integrity during processing. The Masson's trichrome stained light micrographs illustrate that the DDM scaffolds preserved the normal collagen bundling pattern and normal collagen orientation. Reproduced from [141] with permission from Elsevier B.V

a favourable biochemical environment for stem cells. The DDM scaffold loaded with HUCPVCs possessed adequate mechanical properties and showed similar structure and composition to the natural human dermis. The seeded HUCPVCs successfully attached to the fabricated DDM scaffold [141], suggesting that the ECM-like ultrastructure of the scaffold may provide the necessary ligands for cell attachment that facilitate cellular communication between the adjacent cells [141, 144]. Transplanted in vivo, the HUCPVCs-loaded DDM scaffolds enhanced healing of full thickness excisional wounds in diabetic rats by affecting granulation tissue formation and epithelial regeneration [141]. Each tissue type has a specialized ECM structure and functional proteins, glycoproteins, and proteoglycans organized in a 3D system, and the surrounding ECM controls the cells structurally and biochemically. dECM is produced by extracting the ECM of a tissue or organ from the cells within it by chemical, physical, or enzymatic means and preserving the native composition and architecture of the ECM [139]. Therefore, dECM-based bioinks can produce 3D tissues or organs and mimic a complex biological and biochemical microenvironment and a specific ECM composition [139, 145]. The outstanding advantage of these bioinks is that the dECM can be thoroughly recellularized without destroying vital components of the ECM. On the other hand, dECMs can promote cell survival, maturation, differentiation, migration and proliferation [139, 146–148]. Moreover, dECM-based bioinks have a viscoelastic behavior and share rheological properties of native ECMs, including shear viscosity and shear modulus, which strongly depend on the protein composition of the respective dECMs [149]. However, during decellularization, the mechanical properties of the starting tissue are lost, which can have a negative effect on correct cell fate. Compared to other commercial starting materials, the chemical versatility of dECM in cross-linking and post-functionalization is limited. This is reflected in the limited range of studies reporting the 3D printing and/or electrospinning of dECM to date [142]. For skin TE, dECM-based bioinks could be isolated from cell sources related to skin tissue or from stem cells and used to produce 3D skin [139]. Won et al. [150] extracted dECMs from porcine skin using chemical and enzymatic methods to obtain collagen, GAGs, bioactive materials, and GFs in the ECM structure. They used the obtained dECM powder at different concentrations (2 and 3%) as a bioink along with human dermal fibroblasts to fabricate a 3D construct by extrusion-based 3D bioprinting. The main components of ECM (collagen, GAG and elastin) were retained in the dECM-based 3D construct, indicating a positive influence of such a bioink on skin regeneration. This could be due to mimicking the in vivo 3D environment, which results in high viability and proliferation of human skin-derived cells. Furthermore, the used bioink influenced the differentiation of keratinocytes leading to the development of the epidermis through interaction with the dermis [150]. Similarly, Kim and co-workers [151] prepared a porcine skin-derived dECM to develop a printable bioink for a 3D bioprinted skin construct. The proposed bioink largely preserved intrinsic components such as GFs and cytokine proteins that were undetectable in commercially available type I collagen. The mature 3D cell printed skin tissue was stable for two weeks and was characterized by minimal shrinkage, improved epidermal organization, and barrier function. In addition, a in vivo wound healing study [151] showed that the 3D-printed skin patch loaded with endothelial progenitor cells (EPCs) together with adipose-derived stem cells (ADSCs) promoted wound healing, re-epithelialization and neovascularization, as well as blood flow. Despite these benefits, the bioink exhibited insufficient mechanical properties and microenvironment recapitulation [151]. Recently, a biological functional hybrid scaffold based on dECM/gelatin/CS composite with high antibacterial activity for skin TE was presented [152]. The high porosity of the scaffold, which is provided with an interconnected pore structure, stimulated cell growth. In addition, the suitable elastic modulus ($\geq$480 kPa) and degradability ($\geq$80% for 15 days) of the scaffolds provided a stable environment for cell proliferation and a suitable degradation rate that corresponded to the formation of new tissue in the skin TE. In addition, the inclusion of CS conferred good antibacterial activity, as well as water and protein absorption capacity to the scaffold to prevent wound infection and maintain moisture and nutrient balance. In vitro cytocompatibility studies showed that the presence of dECM effectively enhanced cell proliferation [152]. Despite these achievements, some concerns and hurdles must be considered when using dECM-based bioinks for further preclinical and clinical applications. One major concern is the cell-related immunogenicity of xenogenic tissues [139, 151].

4.1.8   Synthetic Polymers

Aliphatic polyesters, the most common type of biodegradable synthetic polymers, are widely used in biomedicine mainly because of their adaptable properties, mechanical strength, processability, and nontoxic degradation products. The main advantage of synthetic biopolymers is that they can be tailored to specific functions and properties and are free from disease transmission [2, 8]. The aliphatic polyesters approved by U.S. Food and Drug Agency (FDA), such as poly(lactic acid) (PLA), poly(glycolic acid) (PGA), poly(ε-caprolactone) (PCL) and their copolymers are widely used in skin TE, wound dressings and drug delivery systems [2]. The degradation of aliphatic polyesters usually occurs by hydrolysis of the ester bonds, producing monomers (e.g., lactic acid and glycolic acid), which can be degraded via the metabolic pathway [2, 8]. However, synthetic polymers lack the natural biological epitopes that interact with cell receptors or adhesion proteins and support cell attachment. Therefore, synthetic hydrogels are usually combined with natural polymers to produce a biofunctional composite material with good mechanical properties and high biocompatibility [153]. In this context, Chandrasekaran and colleagues [154] designed a nanofibrous scaffold of poly(L-lactic acid-co-ε-caprolactone) (PLACL) and gelatin that supported human dermal fibroblast proliferation and collagen deposition under in vitro conditions. Electrospinning was used to create a scaffold morphologically and structurally similar to native ECM. The porous structure, mechanical stability (i.e., tensile strength), and loose margins of the resulting nanofibrous PLACL/gelatin scaffold were favourable for cell infiltration and provided sufficient space for fibroblast ingrowth, resulting in the formation of a dermal equivalent. Special attention was also paid to the tensile properties of the electrospun nanofibrous membranes, as they should facilitate cell proliferation and self-degrade during secretion of a new ECM. Indeed, a very high tensile strength may cause the substrate to remain in the wound bed long after regeneration, thus hindering the development of new tissue, whereas a weaker membrane may lost its structural integrity for the required period of TE [154]. Pure PCL was the material of choice for the fabrication of a nanofiber skin graft constructed in the sandwich-like form: radially aligned nanofiber scaffolds at the bottom, nanofiber scaffolds with squarely arranged microwells and structural cues at the top and microskin tissue islands seeded in microwells in between. This nanofiber skin graft combined the use of TE approaches (the use of scaffolds) with the current "gold standard" in wound treatment (autologous skin micrograft) and was able to confine the microskin tissues in the square-arranged wells while providing nanotopographic cues to the cultured fibroblasts and primary rat skin cells to guide and facilitate their migration in vitro. In addition, the sandwich-type nanofiber scaffolds were able to provide uniform distribution of microskin grafts, increase their uptake rate, and accelerate the re-epithelialization of wounds in a rat skin excision injury model. Also, the empty area in the scaffolds was well suited for exudate drainage in wounds [102]. Recently, a three-layer PCL/gelatin scaffold was designed to provide the proper environment for simultaneous regeneration of the skin's epidermis, dermis, and hypodermis (Fig. 4.8). A co-culture model of keratinocytes and dermal fibroblasts confirmed the efficiency of the scaffold in supporting the proliferation and differentiation of various cell types into organized tissue. The scaffold showed improved and accelerated wound healing in vivo [155].

In another study, Miguel et al. [156] demonstrated the fabrication of an asymmetric 3D skin construct (3D_ SAC) using electrospinning and 3D bioprinting techniques. The layer was mimicked by preparing a dense and interconnected nanofiber composite membrane of PCL and silk serine (PCL_SS), which served as a physical protective barrier for the wound site. The top layer was able to prevent microorganism invasion and exhibit excellent mechanical properties (tensile strength, Young's modulus, and elongation at break) very similar to those of natural human skin. In contrast, the bottom hydrogel layer made of CS and sodium alginate (CS/SA), which was bioprinted layer-by-layer, exhibited adequate porosity, wettability, and biological properties to support cell adhesion, migration, and proliferation. Furthermore, it also inhibited bacterial growth [156].

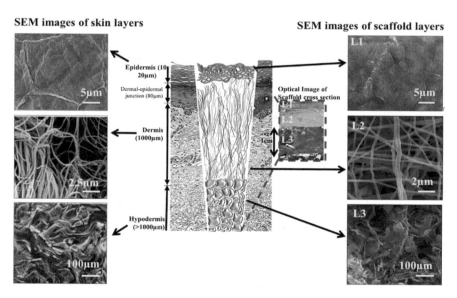

Fig. 4.8 Microstructure of a three-layer PCL/gelatin scaffold. Left: an optical image of the three-layer scaffold, with each layer clearly visible and labelled as L1, L2, and L3. In the middle, a schematic representation of the three-layer scaffold is shown against the background of a skin section stained with hematoxylin–eosin (HE). Corresponding SEM images of the layers can be seen in the right panel. Reproduced from [155] with permission from Elsevier B.V

4.2 Cells

Since most clinical applications require seeded cell culture on scaffold-based models, cells must be reliably obtained and isolated from a reliable lineage that accurately reflects in vivo behaviour. Therefore, cell selection has a remarkable impact on the functionality of the bioengineered skin model. The gold standard cell source in skin TE are autologous cells from patients, which are then propagated in the laboratory to achieve the desired cell count. However, this is an expensive and time-consuming production process, and usually the produced cellular substitutes have a short shelf life [6, 39, 153, 157]. Various cell types can be used instead, including cell lines, primary cells, and pluripotent stem cells or MSCs [6, 8]. Induced pluripotent stem cells (iPSCs) are considered attractive sources because they are versatile, ethically accepted, and genetically tailored to the patient [6, 158]. However, the use of stem cells has other drawbacks, including isolation and characterization of stem cells, insufficient understanding of the mechanisms by which stem cells function, and the tendency of stem cells to self-renew and differentiate, which is strongly influenced by their local environment. Their isolation and characterization is challenging due to their low survival rate and the aseptic conditions required for their cultivation, which requires highly experienced personnel and sophisticated laboratory techniques. In addition, changes in oxygen levels can induce oxidative stress in stem cells, which can affect their phenotype, proliferation rate, and pluripotency. Therefore, extensive differentiation protocols with low degree of stabilization and the possible occurrence of inappropriate phenotypes need to be improved to make stem cells a reliable and trustworthy cell source [38, 159–161]. Classically, fibroblasts and keratinocytes represent in vitro skin, while the inclusion of other cell types (melanocytes, adipocytes, Langerhans cells, nerve cells) are being investigated for future bioengineering of skin equivalents [8, 38, 162]. The incorporation of melanocytes has more recently become common due to the need for pigmented in vitro skin models in the cosmetic industry [38].

4.2.1 Keratinocytes

Keratinocytes are the most abundant cell type studied for most skin tissue constructs. With their keratin deposition, they also contribute to in situ formation of such constructs [39]. They make up 95–97% of the native epidermis and are one of the most accessible cell types in the skin due to their location at the skin surface [8, 162]. Most cell-based skin substitutes, whether scaffold-free models or cellular scaffolds, rely primarily on them for two main reasons: they are easy to produce, and they can serve as a source of stem cells for experimental models and bioengineered therapeutic products [8]. Cultured human keratinocytes can be combined in vitro with dermal substitutes and exposed to air to induce epithelial stratification and cornification. This creates a polarized environment where the culture medium contacts the dermal, and the air contacts the epidermal substitute. Keratinocytes respond to this gradient by orienting proliferating cells toward the medium and cornified cells toward the air to restore the morphology of a stratified squamous epithelium [162]. The ability of the tissue to regenerate itself or undergo abnormal repair (fibrosis) may depend on the local microenvironment provided by the scaffolds in the initial phase of injury/cultivation. The limited information on the combined influence of scaffold microtopography and biomechanical characteristics on wound healing motivated Varkey et al. [163] to study the effects of two different cell populations (keratinocytes and fibroblasts) on substrate properties. An effective way to promote regenerative wound healing versus fibrotic healing could be likely achieved by manipulating the inherent physical properties of the biomaterial-based scaffold, which would represent an easily implementable therapeutic innovation. The biomechanical properties of native ECM are strongly influenced by its specific composition and by post-translation modifications such as cross-linking, transglutamination (modulates collagen maturation and turnover and directs ECM assembly) and glycation (stiffens the collagen matrix by bundling collagen fibres and lowers the surface charge of collagen fibrils), which contribute to intrinsic matrix stiffness, which in turn regulates cell behaviour [163, 164]. Therefore, they investigated the effect of keratinocytes on TE dermal equivalents to determine any biomaterial-mediated antifibrotic influences on engineered skin. Keratinocytes have an inhibitory effect on matrix contraction and stiffness of bioengineered dermal substitutes. This is due to factors released by keratinocytes (transforming growth factor-α, TGF-α, and interleukin (IL)-1), which decreased collagen I production by dermal fibroblasts. Furthermore, keratinocytes also decreased the expression of ECM cross-linking factors, which mediate the cross-linking of collagen and elastin during wound healing. Indeed, the stiffer ECM in fibrotic tissue is thought to be related to extensive collagen cross-linking along with increased collagen deposition. The topography of the scaffold also affects cell behaviour, from cell attachment and migration to differentiation and new ECM synthesis. In addition, the effects of co-culturing dermal equivalents with keratinocytes were investigated based on the changes in pore sizes of the different skin models. Probably due to their effect on collagen synthesis by fibroblasts, keratinocytes also influenced the pore microstructure of the remodeled engineered skin. It was observed that smaller pore sizes were associated with skin models with higher collagen content, which were stiffer. However, further studies are needed to determine whether such differences in pore sizes contribute to altered matrix microtopography that could reduce fibrotic tissue formation [163].

4.2.2 Fibroblasts

The dermis provides structural support to the epidermis and is intertwined with blood vessels and nerves for nutrition and sensation, sweat glands, sebaceous glands, and hair follicles. Fibroblasts are a primary cell type of the dermis. They are responsible for producing the ECM (collagen and elastin fibres) and non-fibrous components such as GAGs and proteoglycans. The dermal region can be divided into an upper papillary and a lower reticular region with their respective types of fibroblasts (papillary or superficial and reticular or deep fibroblasts). The papillary dermis is characterized by thin, randomly aligned collagen fibre bundles arranged in comb-like structures, whereas the reticular dermis consists of numerous thick, properly aligned fibre bundles [5, 8, 165]. The preparation and surface modification of a porous cell scaffold is of the utmost importance in TE. The major problems are seeding of cells with high density, supply of nutrients and oxygen, and cell affinity. To improve cell affinity and solve the problem of cell loss during cell seeding, Yang and colleagues [166] fabricated sponge-like cell scaffolds with different pore structures using blend of porous poly(L-lactic acid) and poly(L-lactic-co-glycolic acid) (PLLA/PLGA). They used a particulate-leaching technique to create porous sponge-like scaffolds and investigated the influence of pore sizes and surface treatment methods with plasma on human skin fibroblasts. The particulate-leaching technique, currently popular in bone TE, involves dissolving the polymer in a suitable solvent and mixing an insoluble salt with the polymer solution. Evaporation of the solvent produces a salt-polymer composite, which is finally washed to remove the salt particles [167]. In Yang's work, pores smaller than 160 μm were reported to be suitable for fibroblast cell growth. In addition, treatment with anhydrous ammonia plasma improved cell affinity and seeding efficiency, which was maintained at over 99% [166]. Developing a structured and organized epithelium requires continuous interactions of epithelial cells with the underlying connective tissue during epidermal regeneration. Engineered skin equivalents can provide an appropriate microenvironment for both fibroblasts and keratinocytes to maintain tissue integrity and achieve tissue function. This ability lies in the in vivo-like dermal matrix, which includes regular basement membrane structures. On this basis, Boehnke and co-workers [168] focused on providing a suitable long-term in vitro system to study the regulatory mechanisms controlling skin regeneration, epidermal tissue homeostasis and stem cell niche establishment in vitro. This study analysed fibroblast growth conditions in a 3D scaffold to optimize the dermal microenvironment by providing an authentic dermal matrix for regular tissue reconstruction and co-cultured keratinocyte function. These skin equivalents showed sustained epidermal viability (over 12 weeks) with regular differentiation, as evidenced by in vivo similar patterns of all differentiation products. Furthermore, the co-cultured keratinocytes exerted a sustained proliferation-stimulating effect on the fibroblasts that populated the scaffold, comparable to a cocktail of fibroblast GFs [168]. To meet the urgent need for skin autografts, Zhang et al. [169] designed acellular dermal matrix (MADM) microcarriers as a culture substrate for fibroblasts and a scaffold for guided tissue regeneration. The developed autografts seem to present a rapid and straightforward strategy for repairing tissue defects. The proposed approach eliminates the need for repeated trypsinization, disrupting cell-ECM interactions and affecting cell viability. This cell expansion protocol simultaneously formed a particulate dermal substitute (EPDS), thus avoiding cell reseeding on the scaffolds. The fabricated MADM retained the ultrastructure of the acellular dermal matrix, exhibited good biocompatibility, and supported the expansion of human fibroblasts either as a direct culture substrate or by culturing cells in conditioned medium prepared from it [169].

4.2.3 Stem Cells

The stratified structure of the epidermis is regularly renewed by epidermal progenitor cells, which reside in the basal layer and in the hair follicles, where skin cells accumulate in the bulge and germ of the hair. Instead of primary skin cells, using other cells in skin substitutes could be beneficial in the complete regeneration of damaged skin [5, 8]. Commonly used cells for skin bioengineering include MSCs and iPSCs [6]. The ability of stem cells (located in microenvironmental niches in the skin) to self-renew is thought to be an essential component of the regenerative capacity of the epidermis [170]. Although autologous skin stem cells are the preferred cell source for 3D skin models and skin regeneration, MSCs derived from other tissues have also been used to improve the functionality of the models. The popularity of the extensive applications of MSCs in skin reconstruction can be attributed to:

(i)   the capacity for self-renewal and multidirectional differentiation,
(ii) the ease of harvesting and low immunogenicity [5, 171].

However, the heterogeneity of MSCs, problems in their isolation, purification and in vitro expansion to differentiated cells limit their commercialization [5]. Adult MSCs have also been used for skin repair and replacement [5, 172, 173]. MSCs are commonly used to enhance the vascularization of artificial skin. They contribute to the vascularization of tissue substitutes through various mechanisms. For example, MSCs, like any other cell type, suffer from hypoxia in scaffolds during the initial phase after implantation. As a result, they release hypoxia-inducible factor-1 (HIF-1)-regulated growth factors that stimulate angiogenesis in the surrounding host tissue. In addition, pluripotent MSCs can differentiate into endothelial cells that are directly integrated into emerging microvascular networks [174]. In recent years, bone marrow-derived MSCs (BM-MSCs), a multipotent stem cell population, have shown promising results in wound healing and skin regeneration [5, 172, 173, 175]. After an injury to the skin, they are mobilized from the bone marrow into the pool of circulating cells. They differentiate into keratinocytes and regulate the proliferation and migration of skin cells to restore the epidermis and promote wound healing in the early inflammatory phase. The benefit of maintaining homeostasis during epidermis development by human BM-MSCs is dose-dependent, i.e., epidermis development is strongly influenced by the percentage of co-cultured BM-MSCs. The BM-MSCs, which were incorporated into the artificial skin models at a low percentage (10%), participated in the basal periphery of the reconstructed epidermis with a similar pattern characteristic of the native epidermis [175]. Positive effects on vascularization of BM-MSCs in small and large animal models were also observed [174]. In addition, promising clinical results have been reported in treating chronic wounds with BM-MSCs seeded on collagen sponges [174, 176]. Recently, MSCs have also been derived from iPSCs and are attractive sources as they can circumvent the limitations of conventional autologous BM-MSCs. iPSC-derived MSCs are rejuvenated during the reprogramming process and have better survival, proliferation and differentiation capacity [6]. These advances in stem cell technology may help provide suitable cell sources for skin bioprinting. Studies show that iPSCs can be differentiated into various types of skin cells capable of forming a multi-differentiated epidermis with hair follicles and sebaceous glands [177].

Adipose-derived MSCs (ADSCs) may be of particular interest for the clinically applicable skin TE. They represent an alternative source of multipotent cells with similar properties to BM-MSCs. In most patients, adipose tissue is available in large quantities and with low morbidity at the donor site, making them a better stem cell source for wound repair and regeneration [174, 178]. The stimulatory effect of ADSCs on cutaneous wound healing may be mediated in part by the paracrine effects of ADSCs on other skin cells [178]. The potential of scaffolds seeded with ADSCs to accelerate wound healing has also been reported. As a cytoprosthetic hybrid to support reconstruction in a rat model of cutaneous wound healing, poly(L-lactide-co-caprolactone)/poloxamer (PLCL/P123) electrospun scaffolds seeded with ADSCs supported the delivery and infiltration of stem cells and their differentiation into epithelial cells [179].

4.2.4 Specialized Cell Types

The major limitation of commonly used skin equivalents is that they can only replicate the epidermal and dermal, and in some cases the hypodermal, layers of the skin. However, skin is a complex organ that comprises more than 50 different cell types, forming other skin components, including the vasculature and immune system, sensory neurons, and appendages (hair follicles, sweat, and sebaceous glands) [180]. In most engineered skin models, it is impossible to reconstruct these macrostructures that are otherwise important regulators of the skin's chemical, physical, and biological functions [153]. Therefore, a fully functional skin substitute requires additional cell components to develop aspects of native physiology that are not limited to ECM deposition but include vascularization, pigmentation, and gland formation. Although stem cells have been influential in regenerative medicine, they are not commonly used for the reconstruction of these skin macrostructures. This is probably due to their shortcomings mentioned above (particularly the possibility of expressing inappropriate phenotypes). The incorporation of melanocytes to achieve desired skin pigmentation and colour, as well as endothelial cells for vascularization, are currently active areas of research. Their successful integration will have a remarkable impact on the outcome of skin substitutes [39].

4.2.4.1 Melanocytes

Melanocytes are pigment-producing cells arising from the neural crest and located in the *stratum basale* layer [181, 182]. Melanin pigments are produced by melanocytes and transferred to neighbouring keratinocytes via their dendrites. The association of one melanocyte and about 40 keratinocytes is called the epidermal melanin unit. The keratinocytes are responsible for the adhesion, proliferation, survival and morphology of melanocytes. The crosstalk between melanocytes and keratinocytes is mediated by direct connections via E-cadherins, which allow the formation of gap junctions and a paracrine effect via keratinocyte-derived soluble factors. Keratinocytes also modulate the transcription of melanogenic proteins and thus the quantity and quality of melanin [183–185]. It took over 30 years to construct a model with functional melanocytes [5, 183]. Co-culture of keratinocytes with melanocytes on a collagen matrix populated with fibroblasts resulted in the proper integration of melanocytes into the epidermal basal layer. Still, melanocytes remained amelanotic (without the characteristic brownish pigment) even after supplementation with promelanogenic factors. Interestingly, normalization of keratinocyte differentiation using keratinocyte growth factor (KGF) instead of epidermal growth factor (EGF) eventually allowed the development of an active pigmentation system. This was demonstrated by the expression of key melanogenic markers and the production and transfer of melanosome-containing melanin into keratinocytes. Moreover, induction of pigmentation was achieved by treatment with known pigmentation modulators, α-melanocyte-stimulating hormone (α-MSH) and forskolin, demonstrating the functionality of the pigmentation system [183]. Keratinocytes and dermal fibroblasts participate in the pigmentation of the human epidermis [186, 187]. It has been presumed that fibroblasts in the dermis regulate the constitutive colour of the various phototypes of the skin and their responses to the environment mainly through the factors they secrete. One of the secreted factors, neuregulin-1 (NRG-1), which is highly expressed by fibroblasts from darker skin, significantly increases pigmentation in a reconstructed skin model and cultured human melanocytes, suggesting its potential role in regulating constitutive human skin colour and possibly its dysfunction in pigmentary skin diseases [186]. Another study also showed that melanocytes in a reconstructed skin model are sensitive to regulation by dermal fibroblasts, basement membrane proteins, and the addition of α-MSH. Moreover, the presence of basement membrane proteins is necessary for the positional orientation of melanocytes, while the addition of α-MSH limits the pigmentation process in the 3D reconstructed skin model [187]. Repigmentation of vitiligo requires an increase in the number and migration of melanocytes into the depigmented epidermis. Various skin grafting or cell transplantation methods have been used to treat this pigmentary disorder. However, the long-term results of cultured melanocyte transplantation have been poorer than those of epidermal grafts. Attempts have been made to transplant melanocytes and keratinocytes simultaneously to increase engraftment rates and simplify procedures because keratinocytes secrete numerous GFs for melanocyte proliferation and migration. Since culturing keratinocytes requires additional skin specimens, it has been suggested that ADSCs could be a potential substitute for keratinocytes in co-cultures with melanocytes. Although the proliferation and migration induced by ADSCs was lower than that of keratinocytes, co-culture of primary melanocytes with ADSCs showed better pigmentation efficacy, including durability and short-term safety, than that of melanocyte culture alone [188].

4.2.4.2   Langerhans Cells

Langerhans cells are a subset of immature dendritic cells located suprabasally in the epidermis. Langerhans cells form a sentinel network in the epidermis, where they take up and process foreign antigens. They undergo an activation process and migrate from the skin to the regional lymph nodes, where they present antigen fragments to native T cells and initiate the cutaneous immune response [189, 190]. Due to their fundamental role in expressing antigens in immunological responses [191], the integration of Langerhans cells into in vitro organotypic constructs has attracted much attention. Similar to melanocytes, the development, recruitment and retention of Langerhans cells in an artificial epidermal layer are regulated by interaction with keratinocytes through several pathways [5]. The cultivation of human primary Langerhans cells is challenging. However, immortalized cell lines such as the CD34+ human acute myeloid leukaemia cell line, MUTZ-3, could differentiate into Langerhans cell-like cells [189, 192]. In general, there are two ways to study them outside the human body, namely skin grafts onto "nude" mice [193] and full-thickness in vitro skin model [189, 190, 194]. The first integration of Langerhans cell-like cells into a well-characterized full-thickness skin model was described by Laubach et al. [189]. They employed the MUTZ-3 cell line to design a relevant research tool to study Langerhans cell biology in vitro, in particular aspects of the interactions between Langerhans cells and other skin cells and their contribution to cutaneous immune responses. To reproduce in vitro the mechanisms previously identified as being involved in Langerhans cell activation and migration in the native human skin, Kosten and colleagues have [194] exploited the unique ability of MUTZ-3 derived Langerhans cells (MUTZ-LC) to exhibit similar phenotypic plasticity to their primary counterparts when integrated into a physiologically relevant skin equivalent model (SE-LC). The skin equivalent consisted of a reconstructed epidermis with primary differentiated keratinocytes and MUTZ-LC on a dermis populated with primary fibroblasts. Using allergens and irritants, it was shown that the MUTZ-LC integrated in SE-LC were fully functional and showed dependence on environmental hazards or damage-associated signals. This model represents a promising tool for a variety of studies (e.g., identification of hazards, study of dendritic cell biology), since readout consisted of physiologically relevant functional changes in Langerhans cell behavior. Moreover, it could represent an alternative method in particular for testing chemical allergens and irritants that are difficult to distinguish from each other [194].

4.2.4.3   Endothelial Cells

Since the epidermis is nourished by passive diffusion of nutrients from the dermal capillaries through the dermal–epidermal junction, an absence of blood circulation for 10–15 days is incompatible with its survival [195]. Therefore, the vascularization of a skin equivalent can improve the longevity of the model by providing nutrients to the cultured cells [5, 174, 195–197]. Because of their dermal origin, human dermal microvascular endothelial cells (HuDMECs) may be particularly advantageous for the pre-vascularization of skin substitutes (Fig. 4.9). Under appropriate culture conditions, HuDMECs form networks of highly organotypic capillaries in 3D skin constructs, which are stabilized by host tissue mural cells after transplantation [197].

In practice, scientists have incorporated vascular channels into the dermal layer by culturing fibroblast cells with human umbilical vein endothelial cells (HUVEC), resulting in the spontaneous formation of blood capillaries within the construct [198].

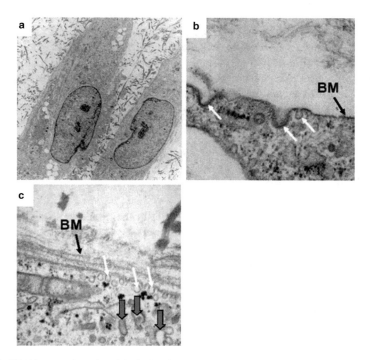

Fig. 4.9 TEM images show the critical role of cell–cell and cell–matrix adhesion in establishing cell polarization and forming the intercellular lumen after immersion in a fibrin gel. **a** HuMECs for 2 days in a fibrin matrix without a formed basement membrane (BM). **b** Seven days after deposition, a BM begins to deposit, and invaginating vesicles (white arrows) can be seen. **c** Numerous invaginating vesicles (white arrows) indicate extensive pinocytosis (cell drinking) 10 days after plating. The deposition of a BM is nearly complete. Vacuoles develop in the cytoplasm (filled grey arrows). Reproduced from [197] with permission from Mary Ann Liebert Inc

Several attempts were also made to create protocols using ADSCs and endothelial progenitor cells (EPCs) in combination to enhance the vascularization process of the skin construct [195, 199]. However, ADSCs only have a significant stimulatory effect on EPCs and not on HUVECs in terms of capillary-like structure formation. On the other hand, the combination of both cell types (ADSCs and EPCs) significantly increased the angiogenic potential of EPCs. This increase in the formation of capillary-like structures by EPCs due to cell–cell contact was associated with significantly increased secretion of vascular endothelial growth factor (VEGF) and VEGF-A mRNA expression [199]. Nevertheless, HUVECs may not be suitable for widespread use in clinical skin TE due to their non-dermal and non-autologous nature. Mature endothelial cells rapidly undergo apoptosis in 3D cultures and are unable to generate proper microvascular networks. For example, microvessels derived from HUVEC monocultures rapidly degenerate. Therefore, the principle of co-cultivation is crucial for the generation of stable and functional microvascular networks [174]. Consistent with this view, Kim et al. [200] have shown that co-transplantation of endothelial cells and smooth muscle cells significantly improves vascularization and healing of skin defects. Alternatively, human blood outgrowth endothelial cells (hBOEC), late outgrowth EPCs derived from human blood, may be the most promising autologous endothelial cell source for future applications in skin TE. Implantation of human dermal fibroblast sheets with hBOECs improved vascularization, epithelial coverage, and matrix organization and prevented excessive contraction of skin defects. This beneficial effect of hBOECs on wound healing is likely mediated by the overlap of GFs and improved oxygen delivery through vascularization [201].

4.3   Growth Factors

The cell microenvironment plays a critical role in the applicability of engineered skin constructs. Part of this microenvironment are GFs responsible for transmitting signals required in the initial stages of cell proliferation and differentiation and ECM secretion [2, 8]. GFs are soluble, secreted signalling polypeptides capable of driving specific cellular responses in a biological environment. Successful tissue growth often depends on their delivery to cells in regenerating tissues. GFs differ from other oligo-/polypeptide molecules (insulin and hormones) in the mode of delivery, and the response triggered. Normally, GFs do not act endocrine but interact with the ECM and cells in a dynamic reciprocal manner. Because of their short half-life and slow diffusion rate, they diffuse short distances through the ECM and act locally on a specific cell subpopulation [202, 203]. However, the bidirectional nature of the interactions between GF and ECM enables the ultimate response of a target cell to a particular soluble GF. Namely, GF regulate the ECM by stimulating cells to increase the production of ECM components or by regulating the production of matrix-degrading proteases and their inhibitors [203]. Owing to their critical role in controlling basic cellular functions and their ability to directly trigger and coordinate tissue regeneration, a wide range of GFs have been tested for various therapeutic applications, including hard and soft tissue regeneration and neovascularization of ischemic tissues [202].

During the natural regeneration of skin tissue after wound formation, a complex and choreographed delivery of numerous bioactive molecules to cells occurs. Various GF, cytokines, and chemokines, including fibroblast growth factor (FGF), keratinocyte growth factor (KGF), IL-1α, and vascular endothelial growth factor (VEGF), are released in the wound bed to trigger cell proliferation, macrophage activation, and angiogenesis [204, 205]. The timing of GF release is precisely controlled and triggered by the progress of the repair. Mimicking endogenous release profiles in an in vitro tissue construct generally requires:

(i)   maintaining the function of proteins, glycoproteins, and other biological molecules during scaffold production,
(ii) precisely controlling the kinetics of GF release,
(iii) potentially controlling two or more types of molecules independently, and
(iv) potentially targeting delivery to specific cell populations [204].

Many studies have confirmed that different types of GF participate in the wound healing process. These include the epidermal growth factor (EGF), FGF, transforming growth factor-β (TGF-β), VEGF, KGF and the platelet-derived growth factor (PDGF). They have a variable influence on the rate of wound healing in acute wounds and can even provide complete healing in chronic wounds [2, 8, 204–206].

4.3.1   Fibroblast Growth Factor

The FGF family comprises 23 members. The three major members involved in skin wound healing are FGF-2, FGF-7, and FGF-10. FGFs are produced by keratinocytes, fibroblasts, endothelial cells, smooth muscle cells, chondrocytes, and mast cells. FGF-2, or basic FGF (bFGF), is elevated in the acute wound and plays a role in granulation tissue formation, re-epithelialization, and tissue remodelling [205]. Specifically, bFGF promotes angiogenesis and regulates many aspects of cellular activity, including cell proliferation, migration, and ECM metabolism, in a time- and concentration-dependent manner [207]. FGF-7 and FGF-10 stimulate the proliferation and migration of keratinocytes, which play a crucial role in re-epithelialization. In addition, FGF-7 and FGF-10 increase transcription factors involved in the detoxification of reactive oxygen species (ROS). This helps to reduce ROS-induced apoptosis of keratinocytes in the wound bed, thereby preserving these cells for re-epithelialization. In addition, FGF-7 may be crucial in the later stages of neovascularization when lumen spaces and basement membranes develop. It is a potent mitogen for vascular endothelial cells and helps to upregulate VEGF. It further stimulates endothelial cells to produce the urokinase-type plasminogen activator, a protease required for neovascularization [205].

A combination of the use of GFs and a porous scaffold could significantly improve the efficacy of skin regeneration, as a porous scaffold alone is not sufficient to induce vascularization and rapid skin regeneration in the early stages of wound healing. With this in mind, biodegradable CS-gelatin microspheres containing bFGF were fabricated and incorporated into a porous CS-gelatin scaffold to enable prolonged, site-specific delivery of bFGF to the grafted skin in a convenient manner. The design of such a scaffold enabled the controlled release of bFGF from CS-gelatin microspheres to human fibroblasts cultured on CS-gelatin scaffolds. Compared to the CS-gelatin scaffolds alone, the scaffolds with bFGF-microspheres significantly augmented cell proliferation, GAG synthesis and mRNA transcription of laminin, which has a promising potential to promote rapid skin regeneration and vascularization [207].

4.3.2 Epidermal Growth Factor

Probably the best-characterised GFs in wound healing are the eEGF family members. In the 1960s, Cohen and Elliot originally described EGF and its "unusual" property of enhancing epidermal keratinization and increasing the overall thickness of the epidermis [208]. EGF is secreted by platelets, macrophages, and fibroblasts and acts on keratinocytes in a paracrine manner. EGF is upregulated after injury and significantly accelerates keratinocyte migration, promotes re-epithelialization, and increases the tensile strength of wounds. Clinical trials in treating chronic wounds show that the addition of topical EGF increases epithelialisation and reduces healing time in skin grafts, venous ulcers and diabetic foot ulcers [205]. An efficient approach to applying EGF locally, maintaining its bioactivity, and protecting the tissue in the reconstruction phase may be the introduction of EGF in a suitable dressing that can release it in a sustained manner and is applied to the wound. Moreover, the new strategies of skin regeneration treatment aim to develop biologically responsive scaffolds capable of delivering multiple bioactive agents and cells to the target tissue. Accordingly, Norouzi and his group [209] fabricated bioactive hybrid nanofibrous scaffolds from gelatin and PLGA encapsulating recombinant human EGF for application in skin TE and as part of wound dressings. The hybrid scaffold showed similar mechanical behaviour to fibrillar collagen and elastin in human skin. The release profile of EGF demonstrated sustained release of the protein in the optimal concentration range required for wound healing. Compared to commercial wound dressings, the gelatin and hybrid scaffolds exhibited improved blood coagulation and platelet adhesion. In addition, cell infiltration into the hybrid scaffold and collagen synthesis were observed [209].

4.3.3   Vascular Endothelial Growth Factor

As can be inferred from its name, VEGF promotes the early events of angiogenesis in wound healing, particularly the migration and proliferation of endothelial cells. An important stimulus for VEGF release in the acute wound is hypoxia due to metabolic disturbances in the wound environment. The resulting angiogenesis restores tissue perfusion and microcirculation. It also increases oxygen tension at the wound site. In animal studies, VEGF administration has been reported to re-establish impaired angiogenesis in diabetic ischemic limbs. In other in vivo experiments, VEGF has been reported to improve the re-epithelialization of diabetic wounds, which is associated with increased vascularization. In addition, VEGF also participates in the inflammatory stages of wound healing by facilitating the recruitment of hematopoietic and inflammatory cells to the wound site and increasing vascular permeability [205].

Biodegradable microspheres can offer precise control of GF release. Cam and colleagues [210] build a similar carrier system based on hydrogels prepared from hyaluronic acid or fibrin and VEGF nanocapsules. After implantation, these VEGF-secreting hydrogels improved wound healing and vessel maturation compared to VEGF-free hydrogel controls (Fig. 4.10). Despite its intrinsic pro-angiogenic activity and contribution to wound homeostasis, fibrin alone did not accelerate wound closure. The natural action of fibrin was enhanced by VEGF nanocapsules, recapitulating the effect of an empty porous HA scaffold [210].

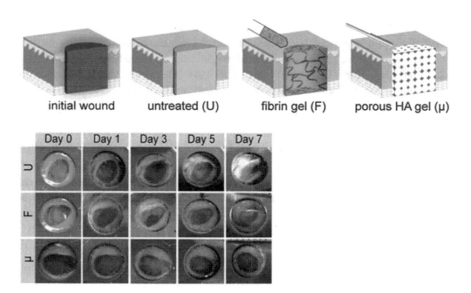

Fig. 4.10   Illustration of untreated (U), fibrin (F), and microporous HA (μ) filled wounds following initial wound creation (top). Digital images of the wounds taken at regular intervals for analysis of wound closure (bottom). Reproduced from [210] with permission from Royal Society of Chemistry

A cost-effective and safe alternative to these microparticles could be a type of fibrin sealant, which can be used as a relatively simple slow-release system within the dermal matrix [174]. Wilcke et al. [211] designed such a slow-release system for VEGF165 and bFGF by incorporating the proteins into a slowly degrading fibrin clot placed in the Integra™ Dermal Regeneration Template. The prolonged protein release proved beneficial for dermal regeneration [174, 211].

4.4   Vascularization

Although artificial skin substitutes have overcome many of the problems associated with skin grafts, their efficacy remains limited because of their inability to deliver blood and nutrients early in wound healing. As a result, cells suffer from oxygen and nutrient deprivation, limiting cell proliferation and, consequently, wound healing and contraction. In addition, incomplete vascularization leads to uneven nutrient and oxygen gradients within scaffolds, leading to a non-uniform cell density as cells survive and proliferate more at scaffold borders. Ultimately, complications caused by inadequate vascularization can cause infection, partial necrosis, delayed healing, and immune responses that result in scaffold rejection after transplantation [212]. Therefore, the lack of sufficient vascularization to support the survival and viability of skin equivalents is the major challenge in developing skin constructs that biologically meet the requirements for skin regeneration [213]. The formation of robust, highly branched, interconnected capillaries that mimic the vascular structure present in vivo is critical for developing a fully functional skin construct. Inadequate vascularization can lead to infection, partial necrosis, or even complete detachment of implanted skin substitutes. Therefore, several strategies have been developed to promote the vascularization of artificial skin. These strategies can be divided into:

(i)   angiogenic: stimulating the ingrowth of blood vessels into implanted tissue constructs,
(ii) pre-vascularizing: generating microvascular networks within tissue constructs before their implantation.

The former is not suitable for the rapid vascularization of large implants due to the slow growth rate of newly formed microvessels. In contrast, inoculation of microvascular networks within tissue constructs with blood vessels at the implantation site leads to rapid blood supply [6, 174, 214].

4.4.1   Scaffold-Based Vascularization Strategies

Dermal scaffolds mimic the natural dermal layer, provide stability, and contain a dense microvascular plexus that nourishes the overlying keratinocytes [215]. Implantation of dermal scaffolds usually induces an angiogenic response with the ingrowth of newly formed microvessels. This process can be further enhanced by modifying the structural and physicochemical properties of the dermal scaffolds [174].

It is noticeable that biomaterials' porosity and other architectural properties play an important role in scaffold revascularization. The pore size and the interconnections between macropores are the main parameters in scaffolds that influence blood vessel growth [216, 217]. The transport of nutrients to the transplanted and host cells through the scaffold is exclusively provided by diffusion processes in the first stage. Since diffusion can only supply the cells with nutrients at a distance <200 μm from the nearest capillary in the surrounding tissue, the transplanted cells in the central part of the scaffold often either fail to grow in or quickly die due to oxygen deficiency, nutrient starvation and insufficient removal of waste products [216]. Thus, the pore size of 200 μm may represent a threshold value to create functional vasculature [212, 218]. Scaffolds with pores >200 μm favour the formation of vascular networks with large blood vessels at low density and deep penetration. This type of scaffold can be used mainly to fabricate large 3D skin substitutes. In contrast, scaffolds with pores <200 μm promote the development of vascular networks with small vessels at high density and low penetration depth [219]. Nevertheless, it seems that the latter is the most suitable for making thin skin substitutes [174]. Although pore size strongly influences vascularization, interconnection of pores is more critical to achieving efficient biomaterial scaffolds. Incomplete pore interconnection could limit the overall biological system and constrain blood vessel invasion. The macropores mainly provide the space for blood vessel growth, while the interconnections function as the gateway for blood vessel ingrowth. Hence, their size determines the area and number of blood vessels that can grow inside the scaffold. It appears that scaffolds with 150 μm interconnections have better vascularization than those with smaller interconnections [217]. However, due to the reticular structure of most dermal scaffolds, differentiation between pore size and interconnection sizes may not be useful in skin TE [174].

The physicochemical properties of the scaffold material markedly influence vascularization. In recent years, surface activation by plasma treatment has been proposed as a promising technique to improve angiogenic tissue response to different types of implants [174, 220]. In addition to modifying commercially available dermal scaffolds, novel scaffold prototypes with enhanced intrinsic bioactivity have been introduced [174]. For example, a hybrid scaffold consisting of a PLGA knitted mesh and a collagen-CS (CCS) scaffold increased the tensile strength of the newly formed skin and its microvessel density after implantation compared with a scaffold consisting of CCS alone. The incorporation of the PLGA knitted mesh into the CCS scaffold improved the mechanical strength with little effect on the mean pore size and porosity, facilitating the process of tissue regeneration and vascularization [221].

An effective strategy to stimulate the ingrowth of blood vessels into scaffolds is incorporating angiogenic GFs, such as VEGF, bFGF, or PDGF. Sustained release of these factors is necessary because free delivery to the wound inevitably leads to rapid loss of GF bioactivity due to protein instability at 37 °C. Another problem GFs of non-human origin lies in their potential immunogenicity [174]. In skin TE, incorporation of GF-loaded microspheres composed of gelatin [222], PLGA [223, 224], or ALG [225] markedly enhanced the angiogenic response of host tissue to dermal scaffolds.

4.4.2 Cell-Based Vascularization Strategies

An implemented approach to improve vascularization in the skin tissue engineering is to seed scaffolds with vasculogenic cells, including different types of mature endothelial cells (HuDMECs, HUVECs), endothelial progenitor cells (EPCs), pluripotent (iPSCs) and MSCs (BM-MSCs, ADSCs) [174]. In addition, pluripotent MSCs can differentiate into endothelial cells that are directly integrated into emerging microvascular networks. It has been reported that the vasculogenic process is significantly enhanced when MSCs are seeded as multicellular spheroids on scaffolds rather than as single cells. Such MSC spheroids are likely to serve as effective vascularization units for skin substitutes by promoting the simultaneous development of new microvascular networks at multiple sites within implants [226].

In addition to isolating purified ADSCs, the stromal vascular fraction (SVF) can also be obtained from adipose tissue and readily be used for a similar purpose. The SVF is a mixture of endothelial cells, stromal cells, and multipotent stem and progenitor cells [174]. Interestingly, these heterogeneous cell populations act synergistically in the rapid formation of mature microvasculature in hydrogel-based dermo-epidermal grafts [227]. Therefore, SVF represents a natural co-culture that can provide optimal conditions for developing functional microvessels in skin constructs. In the future, rapid and effective vascularization could be achieved by integrating adipose tissue-derived microvascular fragments into dermal scaffolds. In contrast to conventional single-cell extractions from adipose tissue, such as purified ADSCs or SVF, microvascular fragments have the distinct advantage of still having functional vascular morphology [174, 228, 229]. Based on the latter, they rapidly reassemble into mature, perfused microvascular networks by establishing interconnections with each other and with the surrounding blood vessels of the host tissue. In addition, the microvascular fragments also contain EPCs and MSCs, which further enhance their enormous angiogenic and regenerative potential [228, 230].

4.4.3 Prevascularized Skin Substitutes

3D bioprinting and microfluidics are rapidly emerging technologies that enable the creation of highly organized microvascular networks with well-defined hierarchical and branching patterns [231, 232]. Endothelial and parenchymal cells, soluble factors, and phase-changing hydrogels can be assembled by 3D bioprinting with high throughput and reproducibility [5, 233]. On the other hand, microfluidic channels in 3D hydrogels can be homogeneously seeded with endothelial cells using appropriate bioreactor perfusion systems. However, while such approaches are already well established in basic research on angiogenesis and microcirculation, their application in skin TE is still in its infancy [174, 234, 235]. To create a functional 3D bioprinted skin equivalent, it is necessary to allow the formation of vascularized constructs and their interconnection by anastomosis with the host vasculature. Chasing this aim, Chen et al. [236] successfully generated a dense microvasculature in 3D hydrogels by encapsulating ECs and human MSCs in a gelatin hydrogel. Recently, Liu and colleagues [237] developed a printable fibrin-based bioink with multiple cell types, including neonatal fibroblasts, pericytes, and human iPSC endothelial cells, which facilitate the fabrication of dermis with self-assembling blood microvessels. The predefined geometry of the 3D bioprinted structure included initial acellular gel regions between bioprinted strips of bioink. Tubulogenesis of endothelial cells within the printed structure was observed approximately 24 h after bioprinting, followed by clear visualization of angiogenesis branching from the printed structure to the cellular regions approximately one week after printing. Vascular bridging between the printed structures became evident after 17 days. The 3D confocal reconstruction images also confirmed the numerous blood vessel formations in the dermis [237]. By combining 3D bioprinting technology with microfluidic principles, Kim et al. [238] proposed a novel platform for fabricating a mature, perfusable vascularized 3D skin equivalent consisting of epidermis, dermis, and hypodermis compartment. The vascularized dermis and hypodermis could promote cross-talks with the epidermis compartment to better recapitulate epidermal morphogenesis by providing a more authentic microenvironment [238]. A new platform for skin irritation was recently engineered by integrating keratinocytes, dermal fibroblasts, and endothelial cells into a microfluidic device. The proposed microfluidic device is equipped with multiple channels and allows the co-cultivation of endothelial cells and keratinocytes with two incompatible media. Compared to other microfluidic devices used to study blood vessels, this platform allowed the seeding of dermal fibroblasts and keratinocytes side by side in the microchannels that mimic the in vivo structure of the skin. In this study, keratinocytes were cultured in a 3D fibrin gel, whereas keratinocytes are normally cultured under 2D conditions to reconstitute the epidermis. Autocrine and paracrine communication between keratinocytes and dermal fibroblasts promoted angiogenesis, vascular sprouting, and collagen deposition under quiescent conditions [239].

Cell sheet engineering is based on cultivating a cell on a thermoresponsive surface or a bioactive silicate material, such as bioactive glass [174, 240, 241]. Scaffold-free generation of artificial skin consisting of multilayered cell sheets is an attractive way to avoid problems related to biocompatibility or cell seeding efficiency typically associated with conventional scaffold-based approaches. In addition, the vascularization capacity of cell sheets can be improved under appropriate culture conditions [174, 241]. In recent years, bioglass has also been used in soft TE (before that, it was the first manufactured inorganic material used in bone TE) because its soluble ions (Ca^{2+}, Si ions) strongly influence cell behaviour [240, 242]. In particular, it stimulates osteogenic differentiation of stem cells and vascularization of endothelial cells through increased secretion of angiogenic GFs from fibroblasts, including VEGF and bFGF. Both GFs participate in the wound healing process, leading Yu and co-workers [240] to activate fibroblast cell sheets in vitro with bioglass. The latter stimulated fibroblasts to secrete several pro-angiogenic GFs. Subsequent implantation of the activated cell sheets in mouse's full-thickness skin defects significantly improved angiogenesis, collagen production, and differentiation of fibroblasts into myofibroblasts at the wound site compared with non-activated control sheets [240]. Cell sheets can also be generated from ADSCs that already have high endogenous pro-angiogenic and vasculogenic capacity [241]. In addition, endothelial cells have been shown to spontaneously form microvascular networks in single-layered and sandwiched cell sheets, demonstrating the potential for engineering prevascularized skin substitutes [174].

4.5   Towards "Authenticity"

To ensure a more accurate representation of human skin, further development of skin equivalents moved towards 3D cell culture and 3D constructs that recapitulate interfaces between cells, cells and matrix, and between cells and the environment, associated with the 3D environment in vivo [38, 39, 238]. From skin biopsies in organ cultures to vascularized organotypic full-thickness reconstructed human skin equivalents, in vitro tissue regeneration of 3D skin has reached a golden era. Nevertheless, 3D skin reconstruction still has room for growth and optimization. The need for reproducible methods, physiological structures, tissue architectures, and blood-perfusable vessels has recently become a reality. However, the requirement of 3D bioprinted skin with a multilayer complex structure of intact skin remains a challenge. Most importantly, the thickness and texture of the epidermal, dermal, and hypodermal layers of the 3D bioprinted skin must be preserved to match the patient's native skin [6, 38, 39, 157]. Various artificial skin models generally consist of an epidermal and a dermal layer, while the hypodermis part of the skin, the hypodermis, is neglected in most current models. Since this skin layer provides the nerves and blood vessels that penetrate the upper layers, the hypodermis plays a key role in re-epithelialization, wound healing, and angiogenesis. In addition, as an endocrine and paracrine organ, the hypodermis plays an important role in tissue irritation and sensitization, as well as drug storage and metabolism [39, 238, 243, 244]. Skin constructs containing multiple functional structures such as sweat glands, hair follicles, and sebaceous glands remain a challenge, although they present an essential requirement [6, 38]. In recent years, the prospect of harnessing the skin microbiome for skin health and personal care has attracted considerable interest. Most of our knowledge of the critical functional role of the microbiota in skin homeostasis comes from mouse models. Still, mouse skin differs from human skin in many ways, including the anatomical structure and gene expression. In this regard, using in vitro human skin models as alternative experimental tools to study the functional mechanisms underlying the cross-talk between the skin and the microbiota is of great interest [246].

4.5.1 Hypodermis Layer

To recapitulate a more realistic microenvironment of native human skin, in vitro skin models should further evolve by including skin-related cells other than the main ones, keratinocytes and fibroblasts. Most of the currently available 3D skin equivalents include an epidermal, dermal or dermo-epidermal layer. However, skin models often dismiss the hypodermis layer as a mere fat storage system. The absence of hypodermis layers limits the application of in vitro skin models, especially concerning the cross-talk of the three skin layers, the regulation of skin morphology, homeostasis and metabolic activity [39, 243, 244, 247, 248]. From a clinical point of view, subdermal adipose tissue is required to heal fourth-degree wounds, where the epidermis, dermis and hypodermis are affected [249]. The increased complexity resulting from the inclusion of a viable adipose layer would allow cell–cell interactions, such as those mediated by factors acting in a paracrine manner. This would improve the physiological relevance of skin equivalents and enable them to better mimic the features of human skin [243, 244, 247]. Many studies have shown that lipophilic adipose tissue, in particular, can absorb harmful substances, indicating a crucial role of hypodermis in evaluating the effects of drugs on the skin and in storing various substances. Therefore, a full-thickness, three-layer skin model that closely mirrors natural skin would likely maintain physiological function over long periods. As such, they would allow for both acute (short-term) and chronic (long-term) studies of skin development and pathogenesis, as well as expand the spectrum of testing to include new test substances and additional endpoints [243, 244]. At first, attempts were made to generate three-layer skin models with stem cells differentiated into the adipogenic lineage [244]. However, stem cells are time-, material-, and cost-intensive. As an alternative, the use of mature adipocytes to create a functional hypodermis is promising. In addition, hypodermis created with stem cells have not had sufficient functional adipocytes. They can be isolated in large numbers and are fully functional without further differentiation from adipose tissue. For skin TE, mature adipocytes can be successfully co-cultured with fibroblasts and keratinocytes to form a functional 3D full skin equivalent [249]. However, accurate reproduction of the different layers and components of human skin is required to prepare in vitro skin models. Simultaneous cultivation of different cell types is a difficult process that needs to be optimized. A frequently cited drawback is the dedifferentiation of mature adipocytes under suboptimal in vitro conditions [243, 249]. By adjusting the culture conditions, it is possible to build a three-layer in vitro skin model that can serve as an alternative to animal experiments [243]. Despite the technical advantage, the maturation of such complex 3D cell-printed constructs, including newly added skin-related cells at the skin tissue level, could be challenging. The main challenges include the difficulty in individually mediating and regulating each cellular region of the complex skin construct in a platform simultaneously [244, 247]. In addition, the required microenvironment in the nutrition and culture period would vary from cellular to tissue maturation in each property of the cell types used. Also, the physicochemical changes due to the addition of other cells, such as shrinkage and deformation, are likely to occur during the maturation process. In light of all this, Kim and colleagues [238] focused on the increasing structural complexity of 3D skin constructs by integrating blood vessels and adipose tissue using 3D cell printing technology. The authors believe that this full-thickness skin model (Fig. 4.11) better reflects the microenvironment than conventional skin constructs, resulting in a more informative and reliable platform for drug screening, cosmetic testing, and basic research [238].

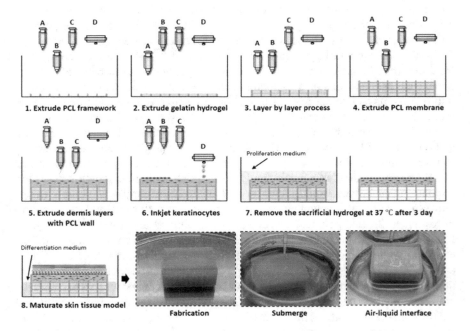

Fig. 4.11 Schematic diagram explaining the cell printing strategy (from side view) for 3D human skin models deposition using a functional transwell system in a single-step process. A: gelatin in the extrusion head. B: PCL in the extrusion head. C: cell-containing collagen bioink in the extrusion head. D: cell-containing medium in the inkjet head. Reproduced from [247] with permission from IOP Publishing

4.5.2 Appendages

One of the problems with in vitro skin models is the lack of skin appendages such as pilosebaceous units, hair follicles, and sweat glands. Due to their absence, those models provide much lower barrier properties than the whole skin. Therefore, the next step towards a genuine physiological skin model is to produce these macrostructures [38, 248, 250, 251]. The in vitro growth of sebocytes [252], eccrine glands [253–256], and hair follicles [248, 257, 258] was tested in a 3D environment [38].

The eccrine sweat glands have the primary function of regulating the human body temperature by evaporating sweat. They are also involved in skin homeostasis, hydration, and immune defence by secreting moisturizing factors such as lactate, urea, and several antimicrobial peptides. They also harbour stem cell populations that can regenerate the epidermis during skin wound healing [248, 253, 254, 259]. Without effective restoration of lost sweat gland cells during wound healing, survivors of extensive deep burns may develop heat intolerance that significantly affects their quality of life [256]. Following injury, the regeneration of sweat glands is of clinical importance but remains largely unresolved due to low regenerative potential and the absence of a clear niche [255]. Several studies on sweat gland development and homeostasis highlight that adult local tissues can ensure epidermal progenitors' behaviour and inductively influence sweat gland specification [259, 260]. These findings have led to the need for additional, more sophisticated in vitro 3D cell culture models to investigate further the process of sweat secretion [254, 255]. One strategy for developing a 3D sweat gland model is producing spheroids by the hanging drop method. This organotypic 3D sweat gland model integrated regulatory and important components relevant to sweat secretion. It also demonstrated physiological regulation of the sweat secretion process, making it ideal for studying unknown sweating mechanisms in vitro [254]. To provide the spatial inductive cues for promoting the specific differentiation of epidermal lineages for sweat gland regeneration, which is crucial for the treatment of deep burns or other wounds, 3D bioprinting technology would be far more suitable [38, 255]. Indeed, conventional scaffold fabrication techniques do not provide complete control over pore morphology, architecture, and reproducibility. Moreover, the porous, biodegradable matrices produced by 3D printing can play an important role in cell function. Another advantage of the 3D bioprinting approach is the control of sweat gland differentiation by simultaneously incorporating living EPCs and inductive factors during fabrication without time constraints and culture conditions [256]. However, the biofabrication of these appendages is still limited [255, 256].

The hair follicle has multiple functions; it serves as an important storage area, plays an important role in skin metabolism, contains multiple stem cell lineages with regenerative capabilities, and is the major penetration route for topically applied substances into the skin [248, 261]. Hair follicles have been shown to provide a long-term reservoir for topically applied substances similar in size to the reservoir of the *stratum corneum* at various body sites. Therefore, hair follicles are considered an important target for drug delivery as they are surrounded by a dense network of blood capillaries

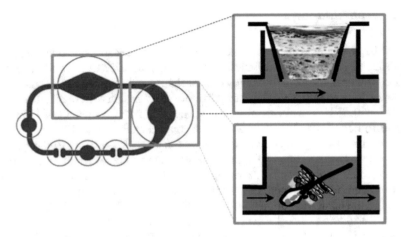

Fig. 4.12 The organ-on-chip application areas are used separately for culturing ex vivo skin biopsies and in vitro skin equivalents in transwells or follicular unit extracts directly in the stream, as shown in the schematic. Reproduced from [248] with permission from Royal Society of Chemistry

and dendritic cells [261]. Considerable progress has been made in developing skin equivalents to model skin as an organ. Still, their static culture limits the replication of essential physiological properties critical for toxicity testing and drug screening. To tackle these limitations, using a dynamically perfused chip-based bioreactor platform could provide variable mechanical shear stress and extend culture duration. Hair follicles cultured in such a bioreactor platform (Fig. 4.12) exhibited a pronounced expansion of hair fibres from the epidermis, indicating an improved simulation of hair follicle biology at the miniature scale. This type of dynamic culture system improved spatial and temporal control of the cellular microenvironment compared to conventional in vitro assays. Moreover, it could be a promising tool for short- and mid-term cultivation and long-term functional organotypic models in the future [248].

Despite advances in culturing and biofabrication techniques, reconstruction of skin appendages in cultures and bioengineered grafts remains a biomedical challenge yet to be overcome [38, 258].

4.5.3   Skin Microbiota

Although the role of the microbiota in skin homeostasis is still emerging, there is growing evidence that an intact microbiota supports the skin barrier. Skin microbiota accomplishes this by providing essential nutrients, influencing our metabolic and hormonal status, displacing potentially pathogenic microbes, strengthening our immune system, and curbing inflammatory responses. The increasing number of research efforts trying to shed more light on the interaction between human skin and microbiota requires the use of appropriate experimental models [246, 262]. In skin biology, reduced diversity and a shift in the microbiota composition, particularly a loss of protective bacteria that limit the outgrowth of pathogenic bacteria, are thought to exacerbate inflammatory skin diseases such as atopic dermatitis (AD) [263, 264]. Most of our understanding of the physiology and biology of the skin comes from mouse models. Nevertheless, they are not an adequate model to study the interaction between skin and microbiota in humans (not to mention the ethical considerations of animal experimentation) [246, 262]. Furthermore, experiments on mice with members of the human microbiota are likely to lead to misleading conclusions. This concern is also supported by the failure to transfer data on *Staphylococcus aureus* (*S. aureus*) from mice to humans, which has further raised scepticism about whether the mouse is an appropriate model for studying skin infections caused by staphylococci [262]. A recent study using a humanized *S. aureus* skin infection model showed marked differences in the inflammatory host mediators involved compared with mouse *S. aureus* infection models. For example, low-dose *S. aureus* inoculation can elicit an IL-8 response in human skin. In contrast, it does not elicit significant inflammation in mouse skin, likely due to langerin (an adhesion factor for *S. aureus* on skin with a remarkable degree of human specificity), highlighting a differential inflammatory response between humans and mice [265]. From this point of view, 3D skin models represent a unique opportunity to functionally analyze the relationship between biomedical and nutritional factors and their influence on the skin microbiota and study the impact of specific host factors on the interaction between skin and microbiota [246, 262].

A humanized *stratum corneum*-based model was used to study skin-derived bacteria in vitro and avoid physiological stress associated with altered environmental conditions. This simple in vitro epidermis model was built on the top layer of the dried human cornea, which served as a substrate and nutrient source for bacterial growth. This *stratum corneum* model has the advantage of allowing the study of growth rates of various skin microbes under much more realistic conditions than the use of conventional agar plates [266]. The major disadvantage of this model is the lack of viable keratinocytes, which prevents the study of the interaction between the microbiota and living keratinocytes. Nevertheless, the *stratum corneum* model has proven useful for studying the interplay between typical abundant members of the microbiota such as *Staphylococcus epidermidis* (*S. epidermidis*) and *Propionibacterium acnes* (*P. acnes*) with potential skin pathogens such as *S. aureus* and *Pseudomonas aeruginosa* [246, 266].

Most studies using 3D human skin models to investigate microbe-skin interactions focus on staphylococci, particularly the common skin resident *S. epidermidis* and the pathogenic transient skin colonizer *S. aureus*. Skin interaction with *S. aureus* has long been a focus of dermatological research due to the high pathogenic potential of *S. aureus*. 3D skin models have proven to be useful tools to study the role of this important skin pathogen, as adherence, growth and localization of *S. aureus* can be easily monitored in such models [246, 262]. By constructing an epidermal skin model based on immortalized keratinocytes with an experimentally introduced filaggrin knock-down, van Drongelen and colleagues [267] demonstrated that diminished filaggrin expression leads to significantly increased epidermal *S. aureus* colonization. This work indicates that genetically engineered immortalized keratinocytes are a valuable tool to study the effects of specific host factors on microbial epidermal colonization in a 3D skin model [246, 267]. Although the role of *S. epidermidis* in skin biology is still emerging, there is growing evidence that *S. epidermidis* plays an important role in skin defence, particularly by directly limiting the growth of pathogenic bacteria such as *S. aureus* [268, 269] and by activating immune cells [270]. In addition to these beneficial attributes, *S. epidermidis* is also a common cause of opportunistic infections, particularly biofilm-associated infections on medical indwelling devices [246]. Thus, there is a great need to increase our knowledge of the interaction between *S. epidermidis* and the skin. Indeed, a growing number of studies using 3D skin models infected with this opportunistic commensal are appearing. For example, a dermal equivalent consisting of a fibrin matrix with viable fibroblasts can be used to monitor the growth of the predominant commensals of the cutaneous microbiota, such as *S. epidermidis*, *P. acnes*, and the fungus *Malassezia furfur*. The use of fibrin as dermal matrix has better supported the development of a fully differentiated epidermis. Primary foreskin-derived keratinocytes were seeded onto the dermal equivalent and a fully differentiated epidermis developed after the keratinocytes were exposed to the air–liquid interface. After applying the previously mentioned commensals, the surface of the skin model was blotted dry with filter paper strips to simulate the dry skin surface in vivo [271].

Non-healing wounds are often associated with bacterial biofilms. A commercial 3D skin equivalent, GraftSkin® (Apligraf®), has been successfully used to monitor biofilm development in wounds infected with *S. aureus* or *P. aeruginosa*, suggesting that such a model could be useful in future studies to evaluate therapeutic approaches to eradicate wound-associated biofilms [272]. Although many published studies using 3D human skin models to investigate the interplay between the skin and the microbiota are still limited, existing studies show tremendous potential for these models in functional microbiota research [246, 262].

References

1. Savoji H, Godau B, Hassani MS, Akbari M (2018) Skin tissue substitutes and biomaterial risk assessment and testing. Front Bioeng Biotechnol 6:86
2. Norouzi M, Boroujeni SM, Omidvarkordshouli N, Soleimani M (2015) Advances in skin regeneration: application of electrospun scaffolds. Adv Healthcare Mater 4(8):1114–1133
3. Chua AWC, Khoo YC, Tan BK, Tan KC, Foo CL, Chong SJ (2016) Skin tissue engineering advances in severe burns: review and therapeutic applications, Burns and trauma 4
4. Korrapati PS, Karthikeyan K, Satish A, Krishnaswamy VR, Venugopal JR, Ramakrishna S (2016) Recent advancements in nanotechnological strategies in selection, design and delivery of biomolecules for skin regeneration. Mater Sci Eng C 67:747–765

5. Yan W-C, Davoodi P, Vijayavenkataraman S, Tian Y, Ng WC, Fuh JY, Robinson KS, Wang C-H (2018) 3D bioprinting XE "3D bioprinting" of skin tissue: from pre-processing to final product evaluation. Adv Drug Deliv Rev 132:270–295

6. Augustine R (2018) Skin bioprinting: a novel approach for creating artificial skin from synthetic and natural building blocks. Prog Biomater 7(2):77–92

7. Dhandayuthapani B, Yoshida Y, Maekawa T, Kumar DS (2011) Polymeric scaffolds in tissue engineering application: a review. Int J Polym Sci 2011

8. Nasalapure AV, Chalannavar RK, Gani RS, Malabadi RB, Kasai DR (2017) Tissue engineering of skin: a review. Trends Biomater Artific Organs 31(2)

9. Rahmani Del Bakhshayesh A, Annabi N, Khalilov R, Akbarzadeh A, Samiei M, Alizadeh E, Alizadeh-Ghodsi M, Davaran S, Montaseri A (2018) Recent advances on biomedical applications of scaffolds in wound healing and dermal tissue engineering. Artific Cells Nanomed Biotechnol 46(4):691–705

10. Nerem R (1992) Tissue engineering in the USA. Med Biol Eng Comput 30(4):CE8–CE12

11. Liaw CY, Ji S, Guvendiren M (2018) Engineering 3D hydrogels for personalized in vitro human tissue models. Adv Healthcare Mater 7(4):1701165

12. Nachman M, Franklin S (2016) Artificial skin model simulating dry and moist in vivo human skin friction and deformation behaviour. Tribol Int 97:431–439

13. Van Kuilenburg J, Masen MA, Van Der Heide E (2013) Contact modelling of human skin: what value to use for the modulus of elasticity? Proc Inst Mech Eng Part J: J Eng Tribol 227(4):349–361

14. Low ZWK, Li Z, Owh C, Chee PL, Ye E, Dan K, Chan SY, Young DJ, Loh XJ (2020) Recent innovations in artificial skin. Biomater Sci 8(3):776–797

15. Han F, Dong Y, Su Z, Yin R, Song A, Li S (2014) Preparation, characteristics and assessment of a novel gelatin–chitosan sponge scaffold as skin tissue engineering material. Int J Pharm 476(1–2):124–133

16. Moore KA, Lemischka IR (2006) Stem cells and their niches. Science 311(5769):1880–1885

17. Tai H, Mather ML, Howard D, Wang W, White LJ, Crowe JA, Morgan SP, Chandra A, Williams DJ, Howdle SM (2007) Control of pore size and structure of tissue engineering scaffolds produced by supercritical fluid processing. Eur Cell Mater 14:64–77

18. Yaniv I, Stein J, Farkas DL, Askenasy N (2006) The tale of early hematopoietic cell seeding in the bone marrow niche. Stem cells and development 15(1):4–16

19. Sekine H, Shimizu T, Sakaguchi K, Dobashi I, Wada M, Yamato M, Kobayashi E, Umezu M, Okano T (2013) In vitro fabrication of functional three-dimensional tissues with perfusable blood vessels. Nat Commun 4(1):1–10

20. Jiankang H, Dichen L, Yaxiong L, Bo Y, Bingheng L, Qin L (2007) Fabrication and characterization of chitosan /gelatin porous scaffolds with predefined internal microstructures. Polymer 48(15):4578–4588

21. Choi YS, Hong SR, Lee YM, Song KW, Park MH, Nam YS (1999) Study on gelatin-containing artificial skin: I. Preparation and characteristics of novel gelatin-alginate sponge. Biomaterials 20(5):409–417

22. Lee SB, Jeon HW, Lee YW, Lee YM, Song KW, Park MH, Nam YS, Ahn HC (2003) Bioartificial skin composed of gelatin and (1→ 3), (1→ 6)-β-glucan. Biomaterials 24(14):2503–2511

23. Harriger MD, Supp AP, Warden GD, Boyce ST (1997) Glutaraldehyde crosslinking of collagen substrates inhibits degradation in skin substitutes grafted to athymic mice. J Biomed Mater Res: Official J Soc Biomater Japanese Soc Biomater 35(2):137–145

24. Tanaka M, Nakakita N, Kuroyanagi Y (1999) Allogeneic cultured dermal substitute composed of spongy collagen containing fibroblasts: evaluation in animal test. J Biomater Sci Polym Ed 10(4):433–453

25. Cooper ML, Hansbrough JF, Spielvogel RL, Cohen R, Bartel RL, Naughton G (1991) In vivo optimization of a living dermal substitute employing cultured human fibroblasts on a biodegradable polyglycolic acid or polyglactin mesh. Biomaterials 12(2):243–248

26. Anisha B, Sankar D, Mohandas A, Chennazhi K, Nair SV, Jayakumar R (2013) Chitosan–hyaluronan/nano chondroitin sulfate ternary composite sponges for medical use. Carbohyd Polym 92(2):1470–1476
27. Cabodi M, Choi NW, Gleghorn JP, Lee CS, Bonassar LJ, Stroock AD (2005) A microfluidic biomaterial. J Am Chem Soc 127(40):13788–13789
28. Li Y, Rodrigues J, Tomas H (2012) Injectable and biodegradable hydrogels: gelation, biodegradation and biomedical applications. Chem Soc Rev 41(6):2193–2221
29. Bryant SJ, Anseth KS (2001) The effects of scaffold thickness on tissue engineered cartilage in photocrosslinked poly (ethylene oxide) hydrogels. Biomaterials 22(6):619–626
30. Tabata Y (2003) Tissue regeneration based on growth factor release. Tissue Eng 9(4, Supplement 1):5–15
31. Vijayavenkataraman S, Lu W, Fuh J (2016) 3D bioprinting of skin: a state-of-the-art review on modelling, materials, and processes. Biofabrication 8(3):032001
32. Parida P, Behera A, Mishra SC (2012) Classification of biomaterials used in medicine. Int J Adv Appl Sci 1:31–35
33. Mathew-Steiner SS, Roy S, Sen CK (2021) Collagen in wound healing. Bioengineering 8(5):63
34. Walimbe T, Panitch A (2020) Best of both hydrogel worlds: harnessing bioactivity and tunability by incorporating glycosaminoglycans in collagen hydrogels. Bioengineering 7(4):156
35. Hunt NC, Grover LM (2010) Cell encapsulation using biopolymer gels for regenerative medicine. Biotech Lett 32(6):733–742
36. Thiele J, Ma Y, Bruekers SM, Ma S, Huck WT (2014) 25th anniversary article: designer hydrogels for cell cultures: a materials selection guide. Adv Mater 26(1):125–148
37. Velegol D, Lanni F (2001) Cell traction forces on soft biomaterials. I. Microrheology of type I collagen gels. Biophys J 81(3):1786–1792
38. Randall MJ, Jüngel A, Rimann M, Wuertz-Kozak K (2018) Advances in the biofabrication of 3D Skin in vitro: healthy and pathological models. Frontiers Bioeng Biotechnol 6:154
39. Yu JR, Navarro J, Coburn JC, Mahadik B, Molnar J, Holmes JH IV, Nam AJ, Fisher JP (2019) Current and future perspectives on skin tissue engineering: key features of biomedical research, translational assessment, and clinical application. Adv Healthcare Mater 8(5):1801471
40. Shoulders MD, Raines RT (2009) Collagen structure and stability. Annu Rev Biochem 78:929–958
41. Kim JE, Kim SH, Jung Y (2016) Current status of three-dimensional printing inks for soft tissue regeneration. Tissue Eng Regenerative Med 13(6):636–646
42. Burke JF, Yannas IV, Quinby WC Jr, Bondoc CC, Jung WK (1981) Successful use of a physiologically acceptable artificial skin in the treatment of extensive burn injury. Ann Surg 194(4):413
43. Bell E, Ehrlich HP, Buttle DJ, Nakatsuji T (1981) Living tissue formed in vitro and accepted as skin-equivalent tissue of full thickness. Science 211(4486):1052–1054
44. Eaglstein WH, Falanga V (1997) Tissue engineering and the development of Apligraf®, a human skin equivalent. Clin Ther 19(5):894–905
45. Hansbrough JF, Boyce ST, Cooper ML, Foreman TJ (1989) Burn wound closure with cultured autologous keratinocytes and fibroblasts attached to a collagen -glycosaminoglycan substrate. JAMA 262(15):2125–2130
46. Huang S, Deng T, Wang Y, Deng Z, He L, Liu S, Yang J, Jin Y (2008) Multifunctional implantable particles for skin tissue regeneration: preparation, characterization, in vitro and in vivo studies. Acta Biomater 4(4):1057–1066
47. Ter Horst B, Chouhan G, Moiemen NS, Grover LM (2018) Advances in keratinocyte delivery in burn wound care. Adv Drug Deliv Rev 123:18–32
48. Chan EC, Kuo S-M, Kong AM, Morrison WA, Dusting GJ, Mitchell GM, Lim SY, Liu G-S (2016) Three dimensional collagen scaffold promotes intrinsic vascularisation for tissue engineering applications. PLoS ONE 11(2):e0149799

49. Choudhury S, Das A (2020) Advances in generation of three-dimensional skin equivalents: pre-clinical studies to clinical therapies. Cytotherapy
50. Charulatha V, Rajaram A (2003) Influence of different crosslinking treatments on the physical properties of collagen membranes. Biomaterials 24(5):759–767
51. Jus S, Stachel I, Schloegl W, Pretzler M, Friess W, Meyer M, Birner-Grünberger R, Guebitz G (2011) Cross-linking of collagen with laccases and tyrosinases. Mater Sci Eng, C 31(5):1068–1077
52. Annabi N, Tamayol A, Uquillas JA, Akbari M, Bertassoni LE, Cha C, Camci-Unal G, Dokmeci MR, Peppas NA, Khademhosseini A (2014) 25th anniversary article: rational design and applications of hydrogels in regenerative medicine. Adv Mater 26(1):85–124
53. Kim YB, Lee H, Kim GH (2016) Strategy to achieve highly porous/biocompatible macroscale cell blocks, using a collagen/genipin-bioink and an optimal 3D printing process. ACS Appl Mater Interfaces 8(47):32230–32240
54. Angele P, Abke J, Kujat R, Faltermeier H, Schumann D, Nerlich M, Kinner B, Englert C, Ruszczak Z, Mehrl R (2004) Influence of different collagen species on physico-chemical properties of crosslinked collagen matrices. Biomaterials 25(14):2831–2841
55. Chau DY, Collighan RJ, Verderio EA, Addy VL, Griffin M (2005) The cellular response to transglutaminase-cross-linked collagen. Biomaterials 26(33):6518–6529
56. Yang Y, Zhu X, Cui W, Li X, Jin Y (2009) Electrospun composite mats of poly [(D, L-lactide)-co-glycolide] and collagen with high porosity as potential scaffolds for skin tissue engineering. Macromol Mater Eng 294(9):611–619
57. Rýglová Š, Braun M, Suchý T (2017) Collagen and its modifications—crucial aspects with concern to its processing and analysis. Macromol Mater Eng 302(6):1600460
58. Hwang Y-J, Larsen J, Krasieva TB, Lyubovitsky JG (2011) Effect of genipin crosslinking on the optical spectral properties and structures of collagen hydrogels. ACS Appl Mater Interfaces 3(7):2579–2584
59. Cornwell KG, Lei P, Andreadis ST, Pins GD (2007) Crosslinking of discrete self-assembled collagen threads: Effects on mechanical strength and cell–matrix interactions. J Biomed Mater Res Part A 80(2):362–371
60. Dong C, Lv Y (2016) Application of collagen scaffold in tissue engineering: recent advances and new perspectives. Polymers 8(2):42
61. Gordon MK, Hahn RA (2010) Collagens. Cell Tissue Res 339(1):247–257
62. Wang TW, Wu HC, Huang YC, Sun JS, Lin FH (2006) Biomimetic bilayered gelatin–chondroitin 6 sulfate-hyaluronic acid biopolymer as a scaffold for skin equivalent tissue engineering. Artif Organs 30(3):141–149
63. Dubský M, Kubinová Š, Širc J, Voska L, Zajíček R, Zajícová A, Lesný P, Jirkovská A, Michálek J, Munzarová M (2012) Nanofibers prepared by needleless electrospinning technology as scaffolds for wound healing. J Mater Sci - Mater Med 23(4):931–941
64. Yoshizato K, Yoshikawa E (1994) Development of bilayered gelatin substrate for bioskin: a new structural framework of the skin composed of porous dermal matrix and thin basement membrane. Mater Sci Eng, C 1(2):95–105
65. Ehrmann A (2021) Non-toxic crosslinking of electrospun gelatin nanofibers for tissue engineering and biomedicine—a Review. Polymers 13(12):1973
66. Hoque ME, Nuge T, Yeow TK, Nordin N, Prasad R (2015) Gelatin based scaffolds for tissue engineering-a review. Polymers Research Journal 9(1):15–32
67. Zhong S, Zhang Y, Lim C (2010) Tissue scaffolds for skin wound healing and dermal reconstruction. Wiley Interdisciplin Rev: Nanomed Nanobiotechnol 2(5):510–525
68. Bello AB, Kim D, Kim D, Park H, Lee S-H (2020) Engineering and functionalization of gelatin biomaterials: from cell culture to medical applications. Tissue Eng Part B Rev 26(2):164–180
69. Biswal T (2021) Biopolymers for tissue engineering applications: a review. Mater Today: Proc 41:397–402
70. Kurian AG, Singh RK, Patel KD, Lee J-H, Kim H-W (2022) Multifunctional GelMA platforms with nanomaterials for advanced tissue therapeutics. Bioactive materials 8:267–295

71. Sisson K, Zhang C, Farach-Carson MC, Chase DB, Rabolt JF (2009) Evaluation of cross-linking methods for electrospun gelatin XE "gelatin" on cell growth and viability. Biomacromol 10(7):1675–1680
72. Kuijpers A, Engbers G, Feijen J, De Smedt S, Meyvis T, Demeester J, Krijgsveld J, Zaat S, Dankert J (1999) Characterization of the network structure of carbodiimide cross-linked gelatin gels. Macromolecules 32(10):3325–3333
73. Zhao X, Lang Q, Yildirimer L, Lin ZY, Cui W, Annabi N, Ng KW, Dokmeci MR, Ghaemmaghami AM, Khademhosseini A (2016) Photocrosslinkable gelatin hydrogel for epidermal tissue engineering. Adv Healthcare Mater 5(1):108–118
74. Bupphathong S, Quiroz C, Huang W, Chung P-F, Tao H-Y, Lin C-H (2022) Gelatin methacrylate hydrogel for tissue engineering applications—a review on material modifications. Pharmaceuticals 15(2):171
75. Sun M, Sun X, Wang Z, Guo S, Yu G, Yang H (2018) Synthesis and properties of gelatin methacryloyl (GelMA) hydrogels and their recent applications in load-bearing tissue. Polymers 10(11):1290
76. Nichol JW, Koshy ST, Bae H, Hwang CM, Yamanlar S, Khademhosseini A (2010) Cell-laden microengineered gelatin methacrylate hydrogels. Biomaterials 31(21):5536–5544
77. Dhandayuthapani B, Krishnan UM, Sethuraman S (2010) Fabrication and characterization of chitosan–gelatin blend nanofibers for skin tissue engineering. J Biomed Mater Res B Appl Biomater 94(1):264–272
78. Abbasian M, Massoumi B, Mohammad-Rezaei R, Samadian H, Jaymand M (2019) Scaffolding polymeric biomaterials: Are naturally occurring biological macromolecules more appropriate for tissue engineering? Int J Biol Macromol 134:673–694
79. Hanumantharao SN, Rao S (2019) Multi-functional electrospun nanofibers from polymer blends for scaffold tissue engineering. Fibers 7(7):66
80. Salerno A, Verdolotti L, Raucci M, Saurina J, Domingo C, Lamanna R, Iozzino V, Lavorgna M (2018) Hybrid gelatin-based porous materials with a tunable multiscale morphology for tissue engineering and drug delivery. Eur Polymer J 99:230–239
81. Zhang Y, Ouyang H, Lim CT, Ramakrishna S, Huang ZM (2005) Electrospinning of gelatin fibers and gelatin/PCL composite fibrous scaffolds. J Biomed Mater Res Part B: Appl Biomater: Official J Soc Biomater Japanese Soc Biomater Australian Soc Biomater Korean Soc Biomater 72(1):156–165
82. Chong EJ, Phan TT, Lim IJ, Zhang Y, Bay BH, Ramakrishna S, Lim CT (2007) Evaluation of electrospun PCL/gelatin nanofibrous scaffold for wound healing and layered dermal reconstitution. Acta Biomater 3(3):321–330
83. Gugerell A, Pasteiner W, Nürnberger S, Kober J, Meinl A, Pfeifer S, Hartinger J, Wolbank S, Goppelt A, Redl H (2014) Thrombin as important factor for cutaneous wound healing: comparison of fibrin biomatrices in vitro and in a rat excisional wound healing model. Wound Repair Regeneration 22(6):740–748
84. Gaule TG, Ajjan RA (2021) Fibrin (ogen) as a therapeutic target: opportunities and challenges. Int J Mol Sci 22(13):6916
85. des Rieux A, Shikanov A, Shea LD (2009) Fibrin hydrogels for non-viral vector delivery in vitro. J Controlled Release 136(2):148–154
86. Storm C, Pastore JJ, MacKintosh FC, Lubensky TC, Janmey PA (2005) Nonlinear elasticity in biological gels. Nature 435(7039):191–194
87. Kopp J, Jeschke MG, Bach AD, Kneser U, Horch RE (2004) Applied tissue engineering in the closure of severe burns and chronic wounds using cultured human autologous keratinocytes in a natural fibrin matrix. Cell Tissue Banking 5(2):89–96
88. Meana A, Iglesias J, Del Rio M, Larcher F, Madrigal B, Fresno M, Martin C, San Roman F, Tevar F (1998) Large surface of cultured human epithelium obtained on a dermal matrix based on live fibroblast-containing fibrin gels. Burns 24(7):621–630.
89. Gorodetsky R, An J, Levdansky L, Vexler A, Berman E, Clark RA, Gailit J, Marx G (1999) Fibrin microbeads (FMB) as biodegradable carriers for culturing cells and for accelerating wound healing. J Investig Dermatol 112(6):866–872

90. Cox S, Cole M, Tawil B (2004) Behavior of human dermal fibroblasts in three-dimensional fibrin clots: dependence on fibrinogen and thrombin concentration. Tissue Eng 10(5–6):942–954

91. Falanga V, Iwamoto S, Chartier M, Yufit T, Butmarc J, Kouttab N, Shrayer D, Carson P (2007) Autologous bone marrow–derived cultured mesenchymal stem cells delivered in a fibrin spray accelerate healing in murine and human cutaneous wounds. Tissue Eng 13(6):1299–1312

92. Raut SY, Gahane A, Joshi MB, Kalthur G, Mutalik S (2019) Nanocomposite clay-polymer microbeads for oral controlled drug delivery: development and in vitro and in vivo evaluations. J Drug Delivery Sci Technol 51:234–243

93. Stark HJ, Boehnke K, Mirancea N, Willhauck MJ, Pavesio A, Fusenig NE, Boukamp P (2006) Epidermal homeostasis in long-term scaffold-enforced skin equivalents. J Investigative Dermatol Symposium Proc, Elsevier, 93–105

94. Jeong K-H, Park D, Lee Y-C (2017) Polymer-based hydrogel scaffolds for skin tissue engineering applications: a mini-review. J Polym Res 24(7):1–10

95. Kogan G, Šoltés L, Stern R, Gemeiner P (2007) Hyaluronic acid: a natural biopolymer with a broad range of biomedical and industrial applications. Biotech Lett 29(1):17–25

96. Owczarczyk-Saczonek A, Zdanowska N, Wygonowska E, Placek W (2021) The immunogenicity of hyaluronic fillers and its consequences. Clin Cosmetic Investigational Dermatol 14:921

97. Alcântara CEP, Noronha MS, Cunha JF, Flores IL, Mesquita RA (2018) Granulomatous reaction to hyaluronic acid filler material in oral and perioral region: a case report and review of literature. J Cosmet Dermatol 17(4):578–583

98. Burdick JA, Prestwich GD (2011) Hyaluronic acid hydrogels for biomedical applications. Adv Mater 23(12):H41–H56

99. Huang G, Chen J (2019) Preparation and applications of hyaluronic acid and its derivatives. Int J Biol Macromol 125:478–484

100. De Angelis B, D'Autilio MFLM, Orlandi F, Pepe G, Garcovich S, Scioli MG, Orlandi A, Cervelli V, Gentile P (2019) Wound healing: in vitro and in vivo evaluation of a biofunctionalized scaffold based on hyaluronic acid and platelet-rich plasma in chronic ulcers. J Clin Med 8(9):1486

101. Anilkumar T, Muhamed J, Jose A, Jyothi A, Mohanan P, Krishnan LK (2011) Advantages of hyaluronic acid as a component of fibrin sheet for care of acute wound. Biologicals 39(2):81–88

102. Hu MS, Maan ZN, Wu J-C, Rennert RC, Hong WX, Lai TS, Cheung AT, Walmsley GG, Chung MT, McArdle A (2014) Tissue engineering and regenerative repair in wound healing. Ann Biomed Eng 42(7):1494–1507

103. Lam PK, Chan ES, To EW, Lau CH, Yen SC, King WW (1999) Development and evaluation of a new composite Laserskin graft. J Trauma Acute Care Surg 47(5):918

104. Uppal R, Ramaswamy GN, Arnold C, Goodband R, Wang Y (2011) Hyaluronic acid nanofiber wound dressing—production, characterization, and in vivo behavior. J Biomed Mater Res B Appl Biomater 97(1):20–29

105. Horch RE, Wagner G, Bannasch H, Kengelbach-Weigand A, Arkudas A, Schmitz M (2019) Keratinocyte monolayers on hyaluronic acid membranes as "upside-down" grafts reconstitute full-thickness wounds. Med Sci Monitor: Int Med J Experiment Clin Res 25:6702

106. Deng R, Fang Y, Shen J, Ou X, Liuyan W, Wan B, Yuan Y, Cheng X, Shu Y, Chen B (2018) Effect of esterified hyaluronic acid as middle ear packing in tympanoplasty for adhesive otitis media. Acta Otolaryngol 138(2):105–109

107. Jayakumar R, Prabaharan M, Kumar PS, Nair S, Tamura H (2011) Biomaterials based on chitin and chitosan in wound dressing applications. Biotechnol Adv 29(3):322–337

108. Sami El-banna F, Mahfouz ME, Leporatti S, El-Kemary M, Hanafy AN (2019) Chitosan as a natural copolymer with unique properties for the development of hydrogels. Appl Sci 9(11):2193

109. Cheung RCF, Ng TB, Wong JH, Chan WY (2015) Chitosan: an update on potential biomedical and pharmaceutical applications. Mar Drugs 13(8):5156–5186

110. Ong S-Y, Wu J, Moochhala SM, Tan M-H, Lu J (2008) Development of a chitosan - based wound dressing with improved hemostatic and antimicrobial properties. Biomaterials 29(32):4323–4332
111. Ma J, Wang H, He B, Chen J (2001) A preliminary in vitro study on the fabrication and tissue engineering applications of a novel chitosan bilayer material as a scaffold of human neofetal dermal fibroblasts. Biomaterials 22(4):331–336
112. Revi D, Paul W, Anilkumar T, Sharma CP (2014) Chitosan scaffold co-cultured with keratinocyte and fibroblast heals full thickness skin wounds in rabbit. J Biomed Mater Res Part A 102(9):3273–3281
113. Sun K-H, Chang Y, Reed NI, Sheppard D (2016) α-Smooth muscle actin is an inconsistent marker of fibroblasts responsible for force-dependent TGFβ activation or collagen production across multiple models of organ fibrosis. Am J Physiol-Lung Cellular Mol Physiol 310(9):L824–L836
114. Biagini G, Bertani A, Muzzarelli R, Damadei A, DiBenedetto G, Belligolli A, Riccotti G, Zucchini C, Rizzoli C (1991) Wound management with N-carboxybutyl chitosan. Biomaterials 12(3):281–286
115. Stone CA, Wright H, Devaraj VS, Clarke T, Powell R (2000) Healing at skin graft donor sites dressed with chitosan. Br J Plast Surg 53(7):601–606
116. Azad AK, Sermsintham N, Chandrkrachang S, Stevens WF (2004) Chitosan membrane as a wound-healing dressing: characterization and clinical application. J Biomed Mater Res Part B: Appl Biomater: Official J Soc Biomater Japanese Soc Biomater Australian Soc Biomater Korean Soc Biomater 69(2):216–222
117. Weng M-H (2008) The effect of protective treatment in reducing pressure ulcers for non-invasive ventilation patients. Intensive Crit Care Nurs 24(5):295–299
118. Guo X, Wang Y, Qin Y, Shen P, Peng Q (2020) Structures, properties and application of alginic acid: a review. Int J Biol Macromol 162:618–628
119. Sanchez-Ballester NM, Bataille B, Soulairol I (2021) Sodium alginate and alginic acid as pharmaceutical excipients for tablet formulation: Structure-function relationship. Carbohyd Polym 270:118399
120. Neves MI, Moroni L, Barrias CC (2020) Modulating alginate hydrogels for improved biological performance as cellular 3D microenvironments. Frontiers Bioeng Biotechnol 8:665
121. Xie Y, Gao P, He F, Zhang C (2022) Application of alginate-based hydrogels in hemostasis. Gels 8(2):109
122. Leung V, Hartwell R, Elizei SS, Yang H, Ghahary A, Ko F (2014) Postelectrospinning modifications for alginate nanofiber-based wound dressings. J Biomed Mater Res B Appl Biomater 102(3):508–515
123. Brenner M, Hilliard C, Peel G, Crispino G, Geraghty R, O'Callaghan G (2015) Management of pediatric skin-graft donor sites: a randomized controlled trial of three wound care products. J Burn Care Res 36(1):159–166
124. Maver T, Mohan T, Gradisnik L, Finsgar M, Kleinschek KS, Maver U (2019) Polysaccharide thin solid films for analgesic drug delivery and growth of human skin cells. Front Chem 7:217
125. Zidarič T, Milojević M, Gradišnik L, Kleinschek KS, Maver U, Maver T (2020) Polysaccharide-based bioink formulation for 3d bioprinting of an in vitro model of the human dermis. Nanomaterials 10(4):733
126. Milojević M, Gradišnik L, Stergar J, Klemen MS, Stožer A, Vesenjak M, Dubrovski PD, Maver T, Mohan T, Kleinschek KS (2019) Development of multifunctional 3D printed bioscaffolds from polysaccharides and NiCu nanoparticles and their application. Appl Surf Sci 488:836–852
127. Diekjurgen D, Grainger DW (2017) Polysaccharide matrices used in 3D in vitro cell culture systems. Biomaterials 141:96–115
128. Maver T, Gradisnik L, Kurecic M, Hribernik S, Smrke DM, Maver U, Kleinschek KS (2017) Layering of different materials to achieve optimal conditions for treatment of painful wounds. Int J Pharm 529(1–2):576–588

129. Maver T, Hribernik S, Mohan T, Smrke DM, Maver U, Stana-Kleinschek K (2015) Functional wound dressing materials with highly tunable drug release properties. RSC Adv 5(95):77873–77884
130. Maver T, Smrke D, Kurečič M, Gradišnik L, Maver U, Kleinschek KS (2018) Combining 3D printing and electrospinning for preparation of pain-relieving wound-dressing materials. J Sol-Gel Sci Technol 88(1):33–48
131. Maver T, Gradisnik L, Kurecic M, Hribernik S, Smrke DM, Maver U, Kleinschek KS (2017) Layering of different materials to achieve optimal conditions for treatment of painful wounds. Int J Pharmaceut 529(1–2):576–588
132. Markstedt K, Mantas A, Tournier I, Avila HM, Hagg D, Gatenholm P (2015) 3D bioprinting human chondrocytes with nanocellulose-alginate bioink for cartilage tissue engineering applications. Biomacromol 16(5):1489–1496
133. Ávila HM, Schwarz S, Rotter N, Gatenholm P (2016) 3D bioprinting of human chondrocyte-laden nanocellulose hydrogels for patient-specific auricular cartilage regeneration. Bioprinting 1:22–35
134. Bazou D, Coakley WT, Hayes AJ, Jackson SK (2008) Long-term viability and proliferation of alginate-encapsulated 3-D HepG2 aggregates formed in an ultrasound trap. Toxicol In vitro 22(5):1321–1331
135. Kwon YJ, Peng CA (2002) Calcium-alginate gel bead cross-linked with gelatin as microcarrier for anchorage-dependent cell culture. Biotechniques 33(1):212–4, 216, 218
136. Kong Y, Xu R, Darabi MA, Zhong W, Luo G, Xing MM, Wu J (2016) Fast and safe fabrication of a free-standing chitosan/alginate nanomembrane to promote stem cell delivery and wound healing. Int J Nanomed 11:2543
137. Rosiak P, Latanska I, Paul P, Sujka W, Kolesinska B (2021) Modification of alginates to modulate their physic-chemical properties and obtain biomaterials with different functional properties. Molecules 26(23):7264
138. Hong S, Sycks D, Chan HF, Lin S, Lopez GP, Guilak F, Leong KW, Zhao X (2015) 3D printing of highly stretchable and tough hydrogels into complex, cellularized structures. Adv Mater 27(27):4035–4040
139. Khoshnood N, Zamanian A (2020) Decellularized extracellular matrix bioinks and their application in skin tissue engineering. Bioprinting e00095
140. Kim BS, Kim H, Gao G, Jang J, Cho D-W (2017) Decellularized extracellular matrix: a step towards the next generation source for bioink manufacturing. Biofabrication 9(3):034104
141. Milan PB, Lotfibakhshaiesh N, Joghataie M, Ai J, Pazouki A, Kaplan D, Kargozar S, Amini N, Hamblin M, Mozafari M (2016) Accelerated wound healing in a diabetic rat model using decellularized dermal matrix and human umbilical cord perivascular cells. Acta Biomater 45:234–246
142. Santschi M, Vernengo A, Eglin D, D'Este M, Wuertz-Kozak K (2019) Decellularized matrix as a building block in bioprinting and electrospinning. Curr Opinion Biomed Engineering 10:116–122
143. Chen C-C, Yu J, Ng H-Y, Lee AK-X, Chen C-C, Chen Y-S, Shie M-Y (2018) The physicochemical properties of decellularized extracellular matrix-coated 3D printed poly (ε-caprolactone) nerve conduits for promoting Schwann cells proliferation and differentiation. Materials 11(9):1665
144. Bondioli E, Fini M, Veronesi F, Giavaresi G, Tschon M, Cenacchi G, Cerasoli S, Giardino R, Melandri D (2014) Development and evaluation of a decellularized membrane from human dermis. J Tissue Eng Regen Med 8(4):325–336
145. Jang J, Kim TG, Kim BS, Kim S-W, Kwon S-M, Cho D-W (2016) Tailoring mechanical properties of decellularized extracellular matrix bioink by vitamin B2-induced photo-crosslinking. Acta Biomater 33:88–95
146. Jung JP, Bhuiyan DB, Ogle BM (2016) Solid organ fabrication: comparison of decellularization to 3D bioprinting. Biomaterials research 20(1):1–11
147. Lee H, Han W, Kim H, Ha D-H, Jang J, Kim BS, Cho D-W (2017) Development of liver decellularized extracellular matrix bioink for three-dimensional cell printing-based liver tissue engineering. Biomacromol 18(4):1229–1237

148. Mohiuddin OA, Campbell B, Poche JN, Thomas-Porch C, Hayes DA, Bunnell BA, Gimble JM (2019) Decellularized adipose tissue: biochemical composition, in vivo analysis and potential clinical applications. Cell Biol Transl Med 6:57–70

149. Pati F, Jang J, Ha D-H, Kim SW, Rhie J-W, Shim J-H, Kim D-H, Cho D-W (2014) Printing three-dimensional tissue analogues with decellularized extracellular matrix bioink. Nat Commun 5(1):1–11

150. Won J-Y, Lee M-H, Kim M-J, Min K-H, Ahn G, Han J-S, Jin S, Yun W-S, Shim J-H (2019) A potential dermal substitute using decellularized dermis extracellular matrix derived bio-ink. Artific Cells Nanomed Biotechnol 47(1):644–649

151. Kim BS, Kwon YW, Kong JS, Park GT, Gao G, Han W, Kim MB, Lee H, Kim JH, Cho DW (2018) 3D cell printing of in vitro stabilized skin model and in vivo pre-vascularized skin patch using tissue-specific extracellular matrix bioink: a step towards advanced skin tissue engineering. Biomaterials 168:38–53

152. Xu J, Fang H, Zheng S, Li L, Jiao Z, Wang H, Nie Y, Liu T, Song K (2021) A biological functional hybrid scaffold based on decellularized extracellular matrix/gelatin/chitosan with high biocompatibility and antibacterial activity for skin tissue engineering. Int J Biol Macromol 187:840–849

153. Przekora A (2020) A concise review on tissue engineered artificial skin grafts for chronic wound treatment: can we reconstruct functional skin tissue in vitro? Cells 9(7):1622

154. Chandrasekaran AR, Venugopal J, Sundarrajan S, Ramakrishna S (2011) Fabrication of a nanofibrous scaffold with improved bioactivity for culture of human dermal fibroblasts for skin regeneration. Biomed Mater 6(1):015001

155. Haldar S, Sharma A, Gupta S, Chauhan S, Roy P, Lahiri D (2019) Bioengineered smart trilayer skin tissue substitute for efficient deep wound healing. Mater Sci Eng: C 105:110140

156. Miguel SP, Cabral CS, Moreira AF, Correia IJ (2019) Production and characterization of a novel asymmetric 3D printed construct aimed for skin tissue regeneration. Colloids Surf, B 181:994–1003

157. Suhail S, Sardashti N, Jaiswal D, Rudraiah S, Misra M, Kumbar SG (2019) Engineered skin tissue equivalents for product evaluation and therapeutic applications. Biotechnol J 14(7):1900022

158. Smith AS, Macadangdang J, Leung W, Laflamme MA, Kim D-H (2017) Human iPSC-derived cardiomyocytes and tissue engineering strategies for disease modeling and drug screening. Biotechnol Adv 35(1):77–94

159. Reddy SHR, Reddy R, Babu NC, Ashok G (2018) Stem-cell therapy and platelet-rich plasma in regenerative medicines: A review on pros and cons of the technologies. J Oral Maxillofac Pathol: JOMFP 22(3):367

160. Riha SM, Maarof M, Fauzi MB (2021) Synergistic effect of biomaterial and stem cell for skin tissue engineering in cutaneous wound healing: a concise review. Polymers 13(10):1546

161. Zakrzewski W, Dobrzyński M, Szymonowicz M, Rybak Z (2019) Stem cells: past, present, and future. Stem Cell Res Ther 10(1):1–22

162. Boyce ST, Warden GD (2002) Principles and practices for treatment of cutaneous wounds with cultured skin substitutes. Am J Surg 183(4):445–456

163. Varkey M, Ding J, Tredget EE, Group WHR (2014) The effect of keratinocytes on the biomechanical characteristics and pore microstructure of tissue engineered skin using deep dermal fibroblasts. Biomaterials 35(36):9591–9598

164. Pfisterer K, Shaw LE, Symmank D, Weninger W (2021) The extracellular matrix in skin inflammation and infection. Frontiers Cell Develop Biol 9

165. Sorrell JM, Caplan AI (2004) Fibroblast heterogeneity: more than skin deep. J Cell Sci 117(5):667–675

166. Yang J, Shi G, Bei J, Wang S, Cao Y, Shang Q, Yang G, Wang W (2002) Fabrication and surface modification of macroporous poly (L-lactic acid) and poly (L-lactic-co-glycolic acid)(70/30) cell scaffolds for human skin fibroblast cell culture. J Biomed Mater Res: Official J Soc Biomater Japan Soc Biomater Aust Soc Biomater Korean Soc Biomater 62(3):438–446

167. Sola A, Bertacchini J, D'Avella D, Anselmi L, Maraldi T, Marmiroli S, Messori M (2019) Development of solvent-casting particulate leaching (SCPL) polymer scaffolds as improved three-dimensional supports to mimic the bone marrow niche. Mater Sci Eng C 96:153–165

168. Boehnke K, Mirancea N, Pavesio A, Fusenig NE, Boukamp P, Stark H-J (2007) Effects of fibroblasts and microenvironment on epidermal regeneration and tissue function in long-term skin equivalents. Eur J Cell Biol 86(11–12):731–746

169. Zhang X, Deng Z, Wang H, Yang Z, Guo W, Li Y, Ma D, Yu C, Zhang Y, Jin Y (2009) Expansion and delivery of human fibroblasts on micronized acellular dermal matrix for skin regeneration. Biomaterials 30(14):2666–2674

170. Markeson D, Pleat JM, Sharpe JR, Harris AL, Seifalian AM, Watt SM (2015) Scarring, stem cells, scaffolds and skin repair. J Tissue Eng Regen Med 9(6):649–668

171. Li M, Zhao Y, Hao H, Han W, Fu X (2015) Mesenchymal stem cell–based therapy for nonhealing wounds: today and tomorrow. Wound Repair and Regeneration 23(4):465–482

172. Ochiai H, Kishi K, Kubota Y, Oka A, Hirata E, Yabuki H, Iso Y, Suzuki H, Umezawa A (2017) Transplanted mesenchymal stem cells are effective for skin regeneration in acute cutaneous wounds of pigs. Regenerative Ther 7:8–16

173. Satoh H, Kishi K, Tanaka T, Kubota Y, Nakajima T, Akasaka Y, Ishii T (2004) Transplanted mesenchymal stem cells are effective for skin regeneration in acute cutaneous wounds. Cell Transplant 13(4):405–412

174. Frueh FS, Menger MD, Lindenblatt N, Giovanoli P, Laschke MW (2016) Current and emerging vascularization strategies in skin tissue engineering. Crit Rev Biotechnol

175. Laco F, Kun M, Weber HJ, Ramakrishna S, Chan CK (2009) The dose effect of human bone marrow-derived mesenchymal stem cells on epidermal development in organotypic co-culture. J Dermatol Sci 55(3):150–160

176. Yoshikawa T, Mitsuno H, Nonaka I, Sen Y, Kawanishi K, Inada Y, Takakura Y, Okuchi K, Nonomura A (2008) Wound therapy by marrow mesenchymal cell transplantation. Plast Reconstr Surg 121(3):860–877

177. Aasen T, Raya A, Barrero MJ, Garreta E, Consiglio A, Gonzalez F, Vassena R, Bilić J, Pekarik V, Tiscornia G (2008) Efficient and rapid generation of induced pluripotent stem cells from human keratinocytes. Nat Biotechnol 26(11):1276–1284

178. Lee SH, Jin SY, Song JS, Seo KK, Cho KH (2012) Paracrine effects of adipose-derived stem cells on keratinocytes and dermal fibroblasts. Ann Dermatol 24(2):136–143

179. Gu J, Liu N, Yang X, Feng Z, Qi F (2014) Adiposed-derived stem cells seeded on PLCL/P123 eletrospun nanofibrous scaffold enhance wound healing. Biomed Mater 9(3):035012

180. Abaci HE, Coffman A, Doucet Y, Chen J, Jacków J, Wang E, Guo Z, Shin JU, Jahoda CA, Christiano AM (2018) Tissue engineering of human hair follicles using a biomimetic developmental approach. Nat Commun 9(1):1–11

181. Cichorek M, Wachulska M, Stasiewicz A, Tymińska A (2013) Skin melanocytes: biology and development. Adv Dermatol Allergol /Postępy Dermatologii I Alergologii 30(1):30

182. Brenner M, Hearing VJ (2009) What are melanocytes really doing all day long…?: from the ViewPoint of a keratinocyte: melanocytes—cells with a secret identity and incomparable abilities. Exp Dermatol 18(9):799

183. Duval C, Chagnoleau C, Pouradier F, Sextius P, Condom E, Bernerd F (2012) Human skin model containing melanocytes: essential role of keratinocyte growth factor for constitutive pigmentation—functional response to α-Melanocyte stimulating hormone and forskolin. Tissue Eng Part C Methods 18(12):947–957

184. Haass NK, Smalley KS, Li L, Herlyn M (2005) Adhesion, migration and communication in melanocytes and melanoma. Pigment Cell Res 18(3):150–159

185. Kuphal S, Bosserhoff AK (2012) E-cadherin cell–cell communication in melanogenesis and during development of malignant melanoma. Arch Biochem Biophys 524(1):43–47

186. Choi W, Wolber R, Gerwat W, Mann T, Batzer J, Smuda C, Liu H, Kolbe L, Hearing VJ (2010) The fibroblast-derived paracrine factor neuregulin-1 has a novel role in regulating the constitutive color and melanocyte function in human skin. J Cell Sci 123(18):3102–3111

187. Hedley SJ, Layton C, Heaton M, Chakrabarty KH, Dawson RA, Gawkrodger DJ, Neil SM (2002) Fibroblasts play a regulatory role in the control of pigmentation in reconstructed human skin from skin types I and II. Pigment Cell Res 15(1):49–56

188. Lim W-S, Kim C-H, Kim J-Y, Do B-R, Kim EJ, Lee A-Y (2014) Adipose-derived stem cells improve efficacy of melanocyte transplantation in animal skin. Biomol Ther 22(4):328

189. Laubach V, Zöller N, Rossberg M, Görg K, Kippenberger S, Bereiter-Hahn J, Kaufmann R, Bernd A (2011) Integration of Langerhans-like cells into a human skin equivalent. Arch Dermatol Res 303(2):135–139

190. Ouwehand K, Spiekstra SW, Waaijman T, Breetveld M, Scheper RJ, de Gruijl TD, Gibbs S (2012) CCL5 and CCL20 mediate immigration of Langerhans cells into the epidermis of full thickness human skin equivalents. Eur J Cell Biol 91(10):765–773

191. Yannas I, Burke J, Orgill D, Skrabut E (1982) Wound tissue can utilize a polymeric template to synthesize a functional extension of skin. Science 215(4529):174–176

192. Santegoets SJ, Masterson AJ, Van Der Sluis PC, Lougheed SM, Fluitsma DM, Van Den Eertwegh AJ, Pinedo HM, Scheper RJ, De Gruijl TD (2006) A CD34+ human cell line model of myeloid dendritic cell differentiation: evidence for a CD14+ CD11b+ Langerhans cell precursor. J Leukoc Biol 80(6):1337–1344

193. Czernielewski J, Demarchez M, Prunieras M (1984) Human Langerhans cells in epidermal cell culture, in vitro skin explants and skin grafts onto "nude" mice. Arch Dermatol Res 276(5):288–292

194. Kosten IJ, Spiekstra SW, de Gruijl TD, Gibbs S (2015) MUTZ-3 derived Langerhans cells in human skin equivalents show differential migration and phenotypic plasticity after allergen or irritant exposure. Toxicol Appl Pharmacol 287(1):35–42

195. Auxenfans C, Lequeux C, Perrusel E, Mojallal A, Kinikoglu B, Damour O (2012) Adipose-derived stem cells (ASCs) as a source of endothelial cells in the reconstruction of endothelialized skin equivalents. J Tissue Eng Regen Med 6(7):512–518

196. Hudon V, Berthod F, Black A, Damour O, Germain L, Auger FA (2003) A tissue-engineered endothelialized dermis to study the modulation of angiogenic and angiostatic molecules on capillary-like tube formation in vitro. Br J Dermatol 148(6):1094–1104

197. Montaño I, Schiestl C, Schneider J, Pontiggia L, Luginbühl J, Biedermann T, Böttcher-Haberzeth S, Braziulis E, Meuli M, Reichmann E (2010) Formation of human capillaries in vitro: the engineering of prevascularized matrices. Tissue Eng Part A 16(1):269–282

198. Black AF, Berthod F, L'Heureux N, Germain L, Auger FA (1998) In vitro reconstruction of a human capillary-like network in a tissue-engineered skin equivalent. FASEB J 12(13):1331–1340

199. Strassburg S, Nienhueser H, Stark GB, Finkenzeller G, Torio-Padron N (2013) Human adipose-derived stem cells enhance the angiogenic potential of endothelial progenitor cells, but not of human umbilical vein endothelial cells. Tissue Eng Part A 19(1–2):166–174

200. Kim KL, Song S-H, Choi K-S, Suh W (2013) Cooperation of endothelial and smooth muscle cells derived from human induced pluripotent stem cells enhances neovascularization in dermal wounds. Tissue Eng Part A 19(21–22):2478–2485

201. Hendrickx B, Verdonck K, Van den Berge S, Dickens S, Eriksson E, Vranckx JJ, Luttun A (2010) Integration of blood outgrowth endothelial cells in dermal fibroblast sheets promotes full thickness wound healing. Stem cells 28(7):1165–1177

202. Lee K, Silva EA, Mooney DJ (2011) Growth factor delivery-based tissue engineering: general approaches and a review of recent developments. J R Soc Interface 8(55):153–170

203. Schultz GS, Wysocki A (2009) Interactions between extracellular matrix and growth factors in wound healing. Wound repair regeneration 17(2):153–162

204. Whitaker M, Quirk R, Howdle S, Shakesheff K (2001) Growth factor release from tissue engineering scaffolds. J Pharm Pharmacol 53(11):1427–1437

205. Barrientos S, Stojadinovic O, Golinko MS, Brem H, Tomic-Canic M (2008) Growth factors and cytokines in wound healing. Wound Repair Regeneration 16(5):585–601

206. Grazul-Bilska AT, Johnson ML, Bilski JJ, Redmer DA, Reynolds LP, Abdullah A, Abdullah KM (2003) Wound healing: the role of growth factors. Drugs Today (Barc) 39(10):787–800

207. Liu H, Fan H, Cui Y, Chen Y, Yao K, Goh JC (2007) Effects of the controlled-released basic fibroblast growth factor from chitosan–gelatin microspheres on human fibroblasts cultured on a chitosan–gelatin scaffold. Biomacromol 8(5):1446–1455

208. Cohen S, Elliott GA (1963) The stimulation of epidermal keratinization by a protein isolated from the submaxillary gland of the mouse. J invest dermatol 40(1):1–5

209. Norouzi M, Shabani I, Ahvaz HH, Soleimani M (2015) PLGA/gelatin hybrid nanofibrous scaffolds encapsulating EGF for skin regeneration. J Biomed Mater Res Part A 103(7):2225–2235

210. Cam C, Zhu S, Truong NF, Scumpia PO, Segura T (2015) Systematic evaluation of natural scaffolds in cutaneous wound healing. J Mater Chem B 3(40):7986–7992

211. Wilcke I, Lohmeyer J, Liu S, Condurache A, Krüger S, Mailänder P, Machens H (2007) VEGF 165 and bFGF protein-based therapy in a slow release system to improve angiogenesis in a bioartificial dermal substitute in vitro and in vivo. Langenbecks Arch Surg 392(3):305–314

212. Amirsadeghi A, Jafari A, Eggermont LJ, Hashemi S-S, Bencherif SA, Khorram M (2020) Vascularization strategies for skin tissue engineering. Biomater Sci 8(15):4073–4094

213. Tavakoli S, Klar AS (2021) Bioengineered skin substitutes: Advances and future trends. Appl Sci 11(4):1493

214. Shahin H, Elmasry M, Steinvall I, Söberg F, El-Serafi A (2020) Vascularization is the next challenge for skin tissue engineering as a solution for burn management. Burns 8

215. MacNeil S (2007) Progress and opportunities for tissue-engineered skin. Nature 445(7130):874–880

216. Bai F, Wang Z, Lu J, Liu J, Chen G, Lv R, Wang J, Lin K, Zhang J, Huang X (2010) The correlation between the internal structure and vascularization of controllable porous bioceramic materials in vivo: a quantitative study. Tissue Eng Part A 16(12):3791–3803

217. Xiao X, Wang W, Liu D, Zhang H, Gao P, Geng L, Yuan Y, Lu J, Wang Z (2015) The promotion of angiogenesis induced by three-dimensional porous beta-tricalcium phosphate scaffold with different interconnection sizes via activation of PI3K/Akt pathways. Sci Rep 5(1):1–11

218. Jain RK, Au P, Tam J, Duda DG, Fukumura D (2005) Engineering vascularized tissue. Nat Biotechnol 23(7):821–823

219. Choi SW, Zhang Y, MacEwan MR, Xia Y (2013) Neovascularization in biodegradable inverse opal scaffolds with uniform and precisely controlled pore sizes. Adv Healthcare Mater 2(1):145–154

220. Ring A, Langer S, Schaffran A, Stricker I, Awakowicz P, Steinau H-U, Hauser J (2010) Enhanced neovascularization of dermis substitutes via low-pressure plasma-mediated surface activation. Burns 36(8):1222–1227

221. Wang X, You C, Hu X, Zheng Y, Li Q, Feng Z, Sun H, Gao C, Han C (2013) The roles of knitted mesh-reinforced collagen–chitosan hybrid scaffold in the one-step repair of full-thickness skin defects in rats. Acta Biomater 9(8):7822–7832

222. Li W, Lan Y, Guo R, Zhang Y, Xue W, Zhang Y (2015) In vitro and in vivo evaluation of a novel collagen/cellulose nanocrystals scaffold for achieving the sustained release of basic fibroblast growth factor. J Biomater Appl 29(6):882–893

223. Cao H, Chen M-M, Liu Y, Liu Y-Y, Huang Y-Q, Wang J-H, Chen J-D, Zhang Q-Q (2015) Fish collagen-based scaffold containing PLGA microspheres for controlled growth factor delivery in skin tissue engineering. Colloids Surf, B 136:1098–1106

224. Wang F, Wang M, She Z, Fan K, Xu C, Chu B, Chen C, Shi S, Tan R (2015) Collagen/chitosan based two-compartment and bi-functional dermal scaffolds for skin regeneration. Mater Sci Eng, C 52:155–162

225. Liu Q, Huang Y, Lan Y, Zuo Q, Li C, Zhang Y, Guo R, Xue W (2017) Acceleration of skin regeneration in full-thickness burns by incorporation of bFGF-loaded alginate microspheres into a CMCS–PVA hydrogel. J Tissue Eng Regen Med 11(5):1562–1573

226. Laschke M, Schank T, Scheuer C, Kleer S, Schuler S, Metzger W, Eglin D, Alini M, Menger M (2013) Three-dimensional spheroids of adipose-derived mesenchymal stem cells are potent initiators of blood vessel formation in porous polyurethane scaffolds. Acta Biomater 9(6):6876–6884

227. Klar AS, Güven S, Zimoch J, Zapiórkowska NA, Biedermann T, Böttcher-Haberzeth S, Meuli-Simmen C, Martin I, Scherberich A, Reichmann E (2016) Characterization of vasculogenic potential of human adipose-derived endothelial cells in a three-dimensional vascularized skin substitute. Pediatr Surg Int 32(1):17–27

228. Laschke MW, Menger MD (2015) Adipose tissue-derived microvascular fragments: natural vascularization units for regenerative medicine. Trends Biotechnol 33(8):442–448

229. Pilia M, McDaniel J, Guda T, Chen X, Rhoads R, Allen RE, Corona B, Rathbone CR (2014) Transplantation and perfusion of microvascular fragments in a rodent model of volumetric muscle loss injury. Army Inst Surg Res Fort Sam Houston Tx

230. McDaniel JS, Pilia M, Ward CL, Pollot BE, Rathbone CR (2014) Characterization and multilineage potential of cells derived from isolated microvascular fragments. J Surg Res 192(1):214–222

231. Hasan A, Paul A, Vrana NE, Zhao X, Memic A, Hwang Y-S, Dokmeci MR, Khademhosseini A (2014) Microfluidic techniques for development of 3D vascularized tissue. Biomaterials 35(26):7308–7325

232. Paulsen S, Miller J (2015) Tissue vascularization through 3D printing: will technology bring us flow? Dev Dyn 244(5):629–640

233. Lee V, Singh G, Trasatti JP, Bjornsson C, Xu X, Tran TN, Yoo S-S, Dai G, Karande P (2014) Design and fabrication of human skin by three-dimensional bioprinting. Tissue Eng Part C Methods 20(6):473–484

234. Sutterby E, Thurgood P, Baratchi S, Khoshmanesh K, Pirogova E (2020) Microfluidic skin-on-a-chip models: toward biomimetic artificial skin. Small 16(39):2002515

235. Risueño I, Valencia L, Jorcano J, Velasco D (2021) Skin-on-a-chip models: general overview and future perspectives. APL bioengineering 5(3):030901

236. Chen YC, Lin RZ, Qi H, Yang Y, Bae H, Melero-Martin JM, Khademhosseini A (2012) Functional human vascular network generated in photocrosslinkable gelatin methacrylate hydrogels. Adv Func Mater 22(10):2027–2039

237. Liu X, Michael S, Bharti K, Ferrer M, Song MJ (2020) A biofabricated vascularized skin model of atopic dermatitis for preclinical studies. Biofabrication 12(3):035002

238. Kim BS, Gao G, Kim JY, Cho DW (2019) 3D cell printing of perfusable vascularized human skin equivalent composed of epidermis, dermis, and hypodermis for better structural recapitulation of native skin. Adv Healthcare Mater 8(7):1801019

239. Jusoh N, Ko J, Jeon NL (2019) Microfluidics-based skin irritation XE "irritation" test using in vitro 3D angiogenesis XE "angiogenesis" platform. APL Bioeng 3(3):036101

240. Yu H, Peng J, Xu Y, Chang J, Li H (2016) Bioglass activated skin tissue engineering constructs for wound healing. ACS Appl Mater Interfaces 8(1):703–715

241. Cerqueira M, Pirraco RP, Santos T, Rodrigues D, Frias A, Martins A, Reis R, Marques A (2013) Human adipose stem cells cell sheet constructs impact epidermal morphogenesis in full-thickness excisional wounds. Biomacromol 14(11):3997–4008

242. Gao L, Zhou Y, Peng J, Xu C, Xu Q, Xing M, Chang J (2019) A novel dual-adhesive and bioactive hydrogel activated by bioglass for wound healing. NPG Asia Mater 11(1):1–11

243. Schmidt FF, Nowakowski S, Kluger PJ (2020) Improvement of a three-layered in vitro skin model for topical application of irritating substances. Frontiers Bioeng Biotechnol 8:388

244. Bellas E, Seiberg M, Garlick J, Kaplan DL (2012) In vitro 3D full-thickness skin-equivalent tissue model using silk and collagen biomaterials. Macromol Biosci 12(12):1627–1636

245. Thornton JF, Gosman A (2004) Skin grafts and skin substitutes. Sel Readings Plast Surg 10(1):1–24

246. Rademacher F, Simanski M, Gläser R, Harder J (2018) Skin microbiota and human 3D skin models. Exp Dermatol 27(5):489–494

247. Kim BS, Lee J-S, Gao G, Cho D-W (2017) Direct 3D cell-printing of human skin with functional transwell system. Biofabrication 9(2):025034

248. Ataç B, Wagner I, Horland R, Lauster R, Marx U, Tonevitsky AG, Azar RP, Lindner G (2013) Skin and hair on-a-chip: in vitro skin models versus ex vivo tissue maintenance with dynamic perfusion. Lab Chip 13(18):3555–3561

249. Huber B, Link A, Linke K, Gehrke SA, Winnefeld M, Kluger PJ (2016) Integration of mature adipocytes to build-up a functional three-layered full-skin equivalent. Tissue Eng Part C Methods 22(8):756–764

250. Brohem CA, da Cardeal LBS, Tiago M, Soengas MS, de Barros SBM, Maria-Engler SS (2011) Artificial skin in perspective: concepts and applications. Pigment Cell Melanoma Res 24(1):35–50

251. Takagi R, Ishimaru J, Sugawara A, Toyoshima K-E, Ishida K, Ogawa M, Sakakibara K, Asakawa K, Kashiwakura A, Oshima M (2016) Bioengineering a 3D integumentary organ system from iPS cells using an in vivo transplantation model. Sci Adv 2(4):e1500887

252. Barrault C, Dichamp I, Garnier J, Pedretti N, Juchaux F, Deguercy A, Agius G, Bernard FX (2012) Immortalized sebocytes can spontaneously differentiate into a sebaceous-like phenotype when cultured as a 3D epithelium. Exp Dermatol 21(4):314–316

253. Poblet E, Jimenez F, Escario-Travesedo E, Hardman J, Hernández-Hernández I, Agudo-Mena JL, Cabrera-Galvan J, Nicu C, Paus R (2018) Eccrine sweat glands associate with the human hair follicle within a defined compartment of dermal white adipose tissue. Br J Dermatol 178(5):1163–1172

254. Klaka P, Grüdl S, Banowski B, Giesen M, Sättler A, Proksch P, Welss T, Förster T (2017) A novel organotypic 3D sweat gland model with physiological functionality. PLoS ONE 12(8):e0182752

255. Huang S, Yao B, Xie J, Fu X (2016) 3D bioprinted extracellular matrix mimics facilitate directed differentiation of epithelial progenitors for sweat gland regeneration. Acta Biomater 32:170–177

256. Liu N, Huang S, Yao B, Xie J, Wu X, Fu X (2016) 3D bioprinting matrices with controlled pore structure and release function guide in vitro self-organization of sweat gland. Sci Rep 6(1):1–8

257. Lee J, Böscke R, Tang P-C, Hartman BH, Heller S, Koehler KR (2018) Hair follicle development in mouse pluripotent stem cell-derived skin organoids. Cell Rep 22(1):242–254

258. Lee J, Rabbani CC, Gao H, Steinhart MR, Woodruff BM, Pflum ZE, Kim A, Heller S, Liu Y, Shipchandler TZ (2020) Hair-bearing human skin generated entirely from pluripotent stem cells. Nature 582(7812):399–404

259. Lu CP, Polak L, Rocha AS, Pasolli HA, Chen S-C, Sharma N, Blanpain C, Fuchs E (2012) Identification of stem cell populations in sweat glands and ducts reveals roles in homeostasis and wound repair. Cell 150(1):136–150

260. Biedermann T, Pontiggia L, Böttcher-Haberzeth S, Tharakan S, Braziulis E, Schiestl C, Meuli M, Reichmann E (2010) Human eccrine sweat gland cells can reconstitute a stratified epidermis. J Investig Dermatol 130(8):1996–2009

261. Lademann J, Knorr F, Richter H, Blume-Peytavi U, Vogt A, Antoniou C, Sterry W, Patzelt A (2008) Hair follicles–an efficient storage and penetration pathway for topically applied substances. Skin Pharmacol Physiol 21(3):150–155

262. Emmert H, Rademacher F, Gläser R, Harder J (2020) Skin microbiota analysis in human 3D skin models—"Free your mice." Exp Dermatol 29(11):1133–1139

263. Williams MR, Gallo RL (2017) Evidence that human skin microbiome dysbiosis promotes atopic dermatitis. J Investig Dermatol 137(12):2460–2461

264. Baldwin HE, Bhatia N, Friedman A, Prunty T, Martin R, Seite S (2017) The role of cutaneous microbiota harmony in maintaining a functional skin barrier. SKIN J Cutan Med 1:s139–s139

265. Schulz A, Jiang L, de Vor L, Ehrström M, Wermeling F, Eidsmo L, Melican K (2019) Neutrophil recruitment to noninvasive MRSA at the stratum corneum of human skin mediates transient colonization. Cell Rep 29(5):1074–1081. e5

266. Van Der Krieken DA, Ederveen TH, Van Hijum SA, Jansen PA, Melchers WJ, Scheepers PT, Schalkwijk J, Zeeuwen PL (2016) An in vitro model for bacterial growth on human stratum corneum. Acta Derm Venereol 96(7):873–879

267. van Drongelen V, Haisma EM, Out-Luiting JJ, Nibbering P, El Ghalbzouri A (2014) Reduced filaggrin expression is accompanied by increased Staphylococcus aureus colonization of epidermal skin models. Clin Exp Allergy 44(12):1515–1524

268. Nakatsuji T, Chen TH, Narala S, Chun KA, Two AM, Yun T, Shafiq F, Kotol PF, Bouslimani A, Melnik AV (2017) Antimicrobials from human skin commensal bacteria protect against Staphylococcus aureus and are deficient in atopic dermatitis. Sci Transl Med 9(378)

269. Otto M (2014) Staphylococcus epidermidis pathogenesis. Springer, Staphylococcus Epidermidis, pp 17–31

270. Naik S, Bouladoux N, Linehan JL, Han S-J, Harrison OJ, Wilhelm C, Conlan S, Himmelfarb S, Byrd AL, Deming C (2015) Commensal–dendritic-cell interaction specifies a unique protective skin immune signature. Nature 520(7545):104–108

271. Holland DB, Bojar RA, Jeremy AH, Ingham E, Holland KT (2008) Microbial colonization of an in vitro model of a tissue engineered human skin equivalent–a novel approach. FEMS Microbiol Lett 279(1):110–115

272. Charles CA, Ricotti CA, Davis SC, Mertz PM, Kirsner RS (2009) Use of tissue-engineered skin to study in vitro biofilm development. Dermatol Surg 35(9):1334–1341

Chapter 5
Commercial Skin Equivalents

In the last decades, regenerative medicine has undergone a significant transformation, largely due to advances in skin TE, driven primarily by large commercial companies [1, 2]. In the 1990s, attempts were made to build various tissue-engineered products based on different biological and non-biological scaffold systems. Commercial products developed during this period, such as Apligraf®, Dermagraft®, Carticel®, and Epicel® (Fig. 5.1), have become well established and are growing in their markets. While many types of engineered tissue products were investigated during the earlier phase of TE development, many of the successful products were from the start related to the skin. They were used to treat burns and chronic wounds (Table 5.1) [2]. These skin substitutes usually consist of allogeneic skin cell populations grown in layers and seeded on scaffolds of ECM proteins. Although this type of approach has been used for decades, the commercialization of this technology advanced in the 2000s, largely due to pressure from the cosmetic industry to develop suitable alternatives to animal testing [1].

5.1 Acellular Skin Substitutes

Acellular skin substitutes, also known as acellular dermal substitutes, function as scaffolds by temporarily mimicking ECM-like support. They promote host cell migration, leading to wound healing by replacing the skin equivalent with endogenous host tissue. They can be biologically active or inert and are derived from natural sources, synthetic, or a combination of both [12, 13]. They have been used to replace soft tissue and treat burns since their introduction in 1994 [14]. The most used commercially available acellular dermal substitutes are natural ECM components and can be divided into:

(i)   decellularized extracellular matrices,
(ii) reconstructed extracellular matrices or biological constructs [15, 16].

T. Zidarič et al., *Function-Oriented Bioengineered Skin Equivalents*, Biobased Polymers, https://doi.org/10.1007/978-3-031-21298-7_5

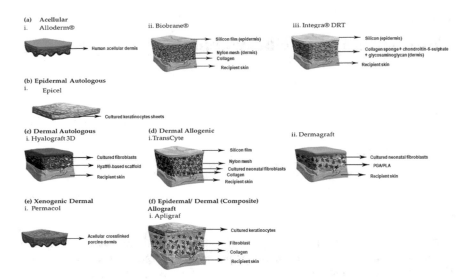

Fig. 5.1 Some of the commercial (acellular and cellular) skin substitutes used for wound treatments. Adapted from [3] with permission from MDPI

The former (decellularized extracellular matrices) are prepared from natural connective tissue (dermis, mesothelium, intestine) of cadaver skin, cleared of all cellular constituents and allergenic/immunogenic substances. The final product is usually freeze-dried to be stored for many months without refrigeration, and it is produced to its final shape without chemical cross-linking [10, 14–16]. The acellular matrix provides a suitable native 3D environment for the influx of fibroblasts and endothelial cells to promote the formation of a new ECM and the vascular network [15]. These materials usually contain portions of the native basement membrane on the papillary side but lack an epidermal layer [15, 16]. The presence of a basement membrane enhances keratinocyte attachment, outgrowth, and differentiation, which is related to laminin and collagen IV in the basement membrane [16]. The most used commercially available decellularized matrices for the treatment of deep wounds after burns and trauma or the reconstruction and treatment of diabetic/venous/pressure ulcers, in general, include: Alloderm®, DermaMatrix™, OASIS®, Permacol™ and EZ-DERM®, which can be placed in the wound bed [10, 14–17].

Reconstructed extracellular matrices are prepared in vitro using the major components of connective tissue (collagen, elastin, GAG, etc.) extracted from animals and then reconstituted. Once ECM components are extracted and purified, they can be combined and processed to form porous 3D structures, such as crosslinking, freeze-drying and electrospinning [15]. In general, the reconstructed extracellular dermis is a synthetic skin substitute because it also contains non-biological molecules and polymers that are not present in normal skin. The artificial nature of these skin substitutes has some distinct advantages and disadvantages compared to natural biological structures. The composition and properties of the product can be controlled

Table 5.1 Commercials products for skin reconstructions/regeneration (available on the market in 2000–2022) (adapted based on [3–11])

Product	Company	Components	Applications	Advantages	Limitations
Alloderm®	LifeCell Inc., Branchburg, NY, USA	Human acellular lyophilized dermis	Partial and full-thickness burns, chronic leg ulcers, diabetic foot ulcers, chronic wounds, plastic reconstruction of surgical wounds	Vascularization of the wound bed, antigenicity, fibroblast repopulation, enhanced release of GFs and cytokines	High cost, risk of disease transmission
Biobrane®	Mylan Bertek Pharmaceuticals, USA	Ultrathin silicone as epidermal analogue film and 3D nylon filament as dermal analogue with type I collagen peptides	Acute full thickness or deep dermal burns, partial and full-thickness wounds, chronic ulcers, traumatic and surgical wounds	Dermal regeneration, exudate drainage, permeable for topical antibiotics, reduced hospitalization and wound healing, pain reduction	Permanent scarring in partial-thickness burn wounds
Integra® Dermal Regeneration Template	Integra® LifeSciences Corp., USA	Dermal analogue—bovine collagen and chondroitin-6-sulfate GAG; epidermal analog—silicone polymer polysiloxane	Acute full thickness or deep dermal burns, partial and full-thickness wounds, chronic ulcers, traumatic and surgical wounds	Dermal regeneration, reduced MMP levels (reduced risk of the onset of chronic nonhealing wound), pain reduction	Low efficiency of graft uptake, risk of wound infection
Epicel®	Genzyme Biosurgery, USA	Sheets of autologous keratinocytes attached to petrolatum gauze support	Superficial, partial, and full-thickness burns, ulcers, chronic wounds	Easy application, permanent coverage of large area wounds	Reduced vascularization, poor mechanical integrity, failure to integrate

(continued)

Table 5.1 (continued)

Product	Company	Components	Applications	Advantages	Limitations
Hyalomatrix PA™	Fidia Advanced Biopolymers, Italy	HYAFF (an ester of hyaluronic acid) layered on silicone membrane	Pressure ulcers, diabetic foot ulcers, traumatic wounds	Generates a richly vascualrized scaffolding for grafting, improved re-epithelization and granulation, improved long-term aesthetic result, reduced risk of immune response	Suitable donor site to collect cells from the patient, prolonged healing time
Hyalograft 3D™	Fidia Advanced Biopolymers, Italy	Cultured fibroblasts hyaluronic acid membrane (HAM)	Deep burns, diabetic foot ulcers	Generates a richly vascualrized scaffolding for grafting, improved re-epithelization and granulation, improved long-term aesthetic result, reduced risk of immune response	Suitable donor site to collect cells from the patient, prolonged healing time
Dermagraft®	Advanced BioHealing, Inc., USA	Bioabsorbable polygalactin mesh matrix seeded with human neonatal fibroblasts and cryopreserved	Partial and full-thickness burns, chronic leg ulcers, painless diabetic wounds, chronic wounds, plastic reconstruction of surgical wounds	Vascularization of the wound bed, antigenicity, fibroblast repopulation, enhanced release of GFs and cytokines, no need to be removed from the wound	High cost

(continued)

Table 5.1 (continued)

Product	Company	Components	Applications	Advantages	Limitations
TransCyte®	Advanced BioHealing, Inc., USA	Collagen-coated nylon mesh seeded with allogenic neonatal human foreskin fibroblasts. Used as a temporary cover	Partial-thickness and deep burns (that do not require autografting), partial and full-thickness wounds	Secretion of the ECM components and GFs, instant access, easy maintenance	High cost, multiple application requirements
Matriderm®	Dr Suwelack Skin and HealthCare AG, Germany	Bovine non-cross-linked lyophilized dermis, coated with a-elastin hydrolysate	Acute full thickness or deep dermal burns, partial and full-thickness wounds, chronic ulcers, traumatic and surgical wounds	Dermal regeneration, reduced MMP levels (reduced risk of the onset of chronic nonhealing wound), pain reduction	Slow vascularisation, high costs, risk of engraftment failure, risk of wound infection
EZ-Derm®	Mölnlycke Health Care AB, Gothenburg, Sweden	Porcine aldehyde cross-linked reconstituted dermal collagen	Acute full thickness or deep dermal burns, partial and full-thickness wounds, chronic ulcers, traumatic and surgical wounds	Dermal regeneration, reduced MMP levels (reduced risk of the onset of chronic nonhealing wounds), pain reduction	No clinically significant benefit over standard options in the treatment of split-thickness wounds
Apligraf®	Organogenesis Inc., USA	Bovine collagen matrix seeded with neonatal foreskin fibroblasts and keratinocytes	Burns, chronic wounds, venous and diabetic foot ulcers, acute surgical defects, skin graft donor sites	Promoting cell migration and proliferation, enhanced release of GFs and cytokines	High cost, short shell-life (range: 5–10 days), risk of disease transmission

(continued)

Table 5.1 (continued)

Product	Company	Components	Applications	Advantages	Limitations
OrCel®	Ortec International Inc., USA	Bovine type I collagen matrix seeded with neonatal foreskin fibroblasts and keratinocytes	Burns, chronic wounds, venous and diabetic foot ulcers, acute surgical defects, skin graft donor sites	Promoting cell migration and proliferation, enhanced release of GFs and cytokines	Risk of disease transmission, few clinical data available that support its use, can not be used for infected wounds
StrataGraft™	Stratatech Corporation, USA	Full thickness skin substitute with dermal and fully differentiated epidermal layers	Burns, chronic wounds, venous and diabetic foot ulcers, acute surgical defects, skin graft donor sites	Promoting cell migration and proliferation, enhanced release of GFs and cytokines	Risk of disease transmission, risk of immune rejection
OASIS® Wound Matrix	Cook Biotech Inc., West Lafayette, IN, USA	Porcine acellular lyophilized small intestine submucosa	Acute full thickness or deep dermal burns, partial and full-thickness wounds, chronic ulcers, traumatic and surgical wounds	Dermal regeneration, reduced MMP levels (reduced risk of the onset of chronic nonhealing wound), pain reduction	Risk of disease transmission, risk of immune rejection

much more precisely. Various additives such as GFs and matrix components can be added to enhance the effect, such as promoting chemotaxis and proliferation of fibroblasts at the wound bed. These products could also avoid complications due to possible disease transmission. However, these synthetic skin substitute products usually lack the basement membrane and its architecture [14]. In this class of acellular extracellular matrices, we find Integra®, Biobrane®, Matriderm®, and Hyalomatrix® [14, 15].

5.1.1 Alloderm®

Alloderm® (LifeCell, New York, NY, USA), an FDA-approved acellular dermal matrix (see Fig. 5.1a) derived from human cadaver skin, has been widely used in various applications since 1992 [10, 15, 17–19]. It is one of the oldest and most widely used decellularized TE skin substitutes permanently integrated into the wound bed. It consists of an ECM in which the cellular components have been removed, leaving the basement membrane intact and the scaffold retaining its collagen and elastic fibre content. These features allow for native tissue ingrowth, revascularization, and remodelling. Decellularization during processing makes AlloDerm® immunologically inert and can be applied without causing a significant immune response [16, 17, 19]. After application, Alloderm® is colonized and vascularized by the underlying cells. However, in some cases, it leads to immunogenic rejection [17, 20]. It is used as a dermal substitute for deep partial- and full-thickness burn wounds and facilitates subsequent autologous split-thickness skin grafting [10]. It is also applied to reconstruct complex surgical defects, such as breast reconstruction [21] and abdominal hernia repair [22]. It has been reported to improve vascularization when used in soft tissues such as the abdominal wall [23, 24]. AlloDerm® can be easily removed, and final reconstruction can then be performed on a healthy bed of granulation tissue after confirmed negative pathological margins [19].

5.1.2 DermaMatrix™

Synthes' alternative to AlloDerm® is DermaMatrix™ (Synthes, West Chester, Pennsylvania, USA) [10]. This material is an allograft derived from donated human skin that can replace, repair, or reinforce soft tissue in grafting procedures such as root coverage and soft tissue ridge augmentation. It is designed to eliminate the need for a second surgical procedure, such as palate harvesting [15]. DermaMatrix™ differs from AlloDerm® in storage, cost, and intraoperative preparation [25]. In a comparative study using an in vivo mouse model, DermaMatrix™ was shown to be superior in maintaining its original shape and consistency three months after subdermal implantation [10, 20]. In another study, results showed that even though DermaMatrix™ is perceived to be thicker and less pliable than AlloDerm®, the skin of patients who underwent breast reconstruction and were treated with DermaMatrix™ expanded faster and obtained a higher final expanded volume. However, this difference did not reach statistical significance [25].

5.1.3   Permacol® and EZ-DERM®

Acellular alternatives derived from animals are also common on the market. Acellular dermal xenografts are often chemically cross-linked, which theoretically makes them less suitable for wound healing. This group includes Permacol™ (Tissue Science Laboratories, Hampshire, UK), a porcine acellular dermal matrix, and EZ-Derm® (Mölnlycke Health Care AB, Gothenburg, Sweden), a porcine-derived dermis collagen matrix [10]. In EZ-DERM®, the collagen was chemically cross-linked with an aldehyde to provide strength and durability [15, 26]. Indeed, the use of Permacol™ (see Fig. 5.1e) as a dermal substitute for wound healing has been largely abandoned [10, 27–30], and the clinical results of EZ-Derm® in wound healing are not convincing [10, 26, 31].

5.1.4   Oasis®

Possibly more suitable for certain forms of wound healing are the, usually not chemically cross-linked, porcine small intestinal submucosa derivatives [10]. OASIS® Wound Matrix (Cook Biotech Inc., West Lafayette, Indianapolis, USA) is a 3D dermal substitute derived from the porcine small intestinal mucosa. It is mainly a collagen-based scaffold, while other ECM components such as GAGs, proteoglycans, fibronectin, and GFs are also part of it. It is available in different sizes as a single or three-layer matrix [10–12, 32]. Based on the fact that it is considered medically necessary to treat chronic, noninfected, partial- or full-thickness vascular ulcers of the lower extremities that have not responded adequately after one month of conventional ulcer therapy [10, 15], OASIS® was also tested in such a study. Namely, a randomized, controlled study of 120 patients with chronic venous leg ulcers showed that significantly more wounds healed (55% versus 34%) when OASIS® Wound Matrix was combined with compression therapy [33]. The FDA classifies it as a class II (moderate risk) product [15].

5.1.5   Integra®

Integra® Dermal Regeneration Template (Integra LifeSciences, Princeton, New York, USA), consisting of bovine collagen, chondroitin-6-sulphate and a silastic (a portmanteau of 'silicone' and 'plastic') membrane (see Fig. 5.1a). This product is widely used in the clinical treatment of deep burn wounds (partial and complete), skin defects of various etiologies, chronic wounds, and soft tissue defects. The bovine collagen dermal analogue integrates with the patient's own cells, and the temporary epidermal silicone is peeled off while the dermis regenerates. A very thin autograft is then grafted onto the neo-dermis [16, 34, 35]. However, with regard to wound infections and graft uptake, Integra® did not achieve such a good result [14, 15].

5.1.6 Matriderm®

Matriderm® (MedSkin Solution Dr. Suwelack, Billerbeck, Germany) is an engineered dermal template designed to repair skin defects in one step. It is a structurally intact matrix of bovine type I collagen with elastin. It is used for dermal regeneration. Its indications are burn wounds of full thickness or deep dermis and chronic wounds. The matrix serves as a support structure for the growth of cells and blood vessels. The elastin component improves the stability and elasticity of the regenerating tissue and provides faster vascularization compared to Integra®. As the healing process progresses, the fibroblasts secrete the ECM and replace the Matriderm®. However, unlike Integra®, Matriderm® has been shown to allow immediate split skin grafting without adverse effects. In experimental models, the matrix reduced wound contracture, and histologically, collagen bundles are more irregularly arranged in the scar. Clinical trials with long-term clinical evaluation showed no difference in scar elasticity between the described dermal substitute and split-thickness skin grafts alone [14, 15, 36].

5.1.7 Biobrane®

Biobrane® (Smith and Nephew, Watford, UK) is composed of bovine type 1 collagen, silicone and nylon and is mechanically bonded to a flexible mesh nylon fabric (see Fig. 5.1a). The semi-permeable membrane is comparable to the human epidermis and controls the loss of water vapour, allows exudate drainage, and provides permeability to topical antibiotics. The nylon/silicone membrane provides flexible, adherent coverage of the wound surface [3, 14–16, 37]. It has proven effective for clean superficial burns and donor sites. The mesh will adhere to the wound until healing occurs underneath when used to cover partial-thickness wounds. Biobrane® should be removed from any full thickness wound before skin grafting. The nylon contained in Biobrane® is not incorporated, making this acellular matrix a wound dressing and not a skin substitute. Biobrane® is a proven synthetic dressing for burn wounds; patients with shallow thickness superficial burns experience less pain than gauze and silver sulfadiazine dressing. Biobrane® also significantly reduces hospitalization, wound healing time and the need for analgesics. However, it has been associated with permanent scarring in partial-thickness burn wounds [14, 15].

5.1.8 Hyalomatrix®

Hyalomatrix® (Fidia Advanced Biopolymers, Abano Terme, Italy) is a bilayer, non-woven pad consisting of a wound contact layer of a derivative of hyaluronic acid ester (Hyaff®) in fibrous form and an outer layer of a semi-permeable silicone membrane. The wound contact layer (HA fibre) is a biodegradable matrix and acts as a 3D scaffold for cellular invasion and capillary growth. It can be colonized by fibroblasts and enables the regular application of ECM components, allowing for orderly reconstruction of the dermal tissue. At the same time, the semi-permeable silicone layer keeps microbes out but gives the skin the air it needs for healing; it also preserves moisture in the wound. In addition, the wound healing process can be observed from the outside through the transparent silicone layer without causing pain. It is specially designed to treat deep burns and full-thickness wounds and serves as a wound preparation for the implantation of autologous skin grafts [3, 15, 16, 38].

5.2 Cellular Skin Substitutes

Cellular skin substitutes contain functional cells embedded in an ECM. These cells can secrete cytokines, GFs, collagen, fibronectin and GAGs that promote angiogenesis, granulation and re-epithelialization [12, 13]. They can be further subdivided into:

(i)   epithelial sheets,
(ii) dermis equivalents, and
(iii) full-thickness (FT) or composite skin equivalents.

Epithelial sheets, such as CEAs and RHEs, are formed by epithelial cells embedded in or seeded onto polymer membranes [15]. Cultured epithelial autografts (CEAs) have become available as an alternative to the use of expanded skin autografts and regrafting [12, 39]. They are typically used for deep dermal or full-thickness burns greater than or equal to 30% of total body surface area. The process of culturing the biopsy taken from the patient allows a relatively small donor site to be expanded into a graft that can cover a large body surface area. If the use of the graft is not immediately required, the cultured autograft can be cryopreserved for later use. Once the expanded cultured autograft is obtained, the graft should be arranged with the cell sheet facing down to maintain the basal–apical orientation of the keratinocyte sheets [12]. The disadvantages of CEAs include the high cost of the graft, the susceptibility of keratinocyte sheets to infection due to degradation by bacterial proteases and cytotoxins, variable graft uptake, and the long time required to culture and produce the epidermal autograft [12, 40]. Analogous to CEAs in commercial applications, models of the reconstructed human epidermis (RHE) are formed from a layer of an inert filter substrate containing collagen with fibroblast cells cultured in an appropriate medium [41, 42]. Usually, they consist only of keratinocytes [41].

Improved methods for their preparation include using melanocytes to obtain a reconstructed epidermis corresponding to the in vivo epidermis [42, 43]. Their advantage is that they have a higher degree of possible standardization [44]. These models are mainly used for skin irritation, skin corrosion, phototoxicity, epidermal genotoxicity, transdermal drug delivery, skin sensitization, and metabolism studies [41–45]. In contrast, RHE models with melanocytes are used for skin lightening and pigmentation [42–44]. Although these reconstructed epidermis models have similar skin properties and can be used in preliminary toxicity and permeation studies, they still have limitations. The major limitation is their relatively weak barrier function, which leads to increased permeability and thus inaccurate transport results in comparison with the native skin [46]. Epidermis models, both CEAs and RHEs, include Epicel®, SkinEthic™, EpiSkin™ and EpiDerm™.

Although most commercial dermal skin substitutes are acellular and promote infiltration of cells and blood vessels from the host tissue, the addition of fibroblasts to the scaffold improves the regulation of regenerative mechanisms [47, 48]. In general, the equivalents of dermis consist of porous 3D matrices or hydrogels containing fibroblasts and provide for the synthesis of dermal ECM components (mainly collagens and elastin). In addition, fibroblasts secrete a variety of GFs and cytokines that have proangiogenic and proinflammatory effects. Allogeneic fibroblasts may pose a potential risk for immune reactions and subsequent rejection. In contrast, neonatal foreskin fibroblasts have underdeveloped human leukocyte antigens, with no observed clinical signs of rejection [49, 50]. However, cell viability after transplantation is reported to be a few weeks [48]. The best-known representatives of this class are Dermagraft®, Transcyte™ and Hyalograft3D™ [12, 15, 51].

Full-thickness skin equivalents are engineered tissues that contain both epithelial and dermal tissue. They consist of epithelial tissue growing on a dermis surrogate composed of fibroblasts encased in a biomaterial [15, 37, 44]. These models, including Apligraf®, EpiDermFT™, and StrataTest®, are typically used to test percutaneous absorption, wound healing, and bacterial adhesion. The advantage of these models is that they are standardized. However, their drawback is their cost and inability to maintain a long-term cell culture [44, 52]. In addition, full-thickness models enriched with melanocytes, Langerhans cells, endothelial cells, and hypodermis models are expanding their use in research and applications. Some examples of their use include the study of melanogenic proteins and the assessment of vitiligo pathogenesis, the assessment of allergens and the study of Langerhans cells maturation and migration. Another similar study is related to angiostatic therapies and lipid metabolism assessment [44, 52]. A major drawback of these models is their lack of validated and standardized commercial availability and their technically demanding nature.

Full-thickness models with stem cells are used to assess epidermis development, wound healing studies, pigmentation disorders, penetration of substances, and enable autologous transplantation. The feasibility of such models has yet to be demonstrated and requires significant preparatory steps and a differentiated protocol for use. Hair follicles can be included simultaneously in full-thickness models and are important for product penetration studies. Still, they have lower throughput and are generally not considered along with other skin components [44, 52].

5.2.1 EpiCel™

EpiCel™ (Genzyme Biosurgery, USA) is a CEA derived from human keratinocytes (see Fig. 5.1b). It is prepared by growing a sheet of autologous keratinocytes two to eight cells thick on mouse 3T3 fibroblasts (for approximately 16 days), followed by attaching the sheet of keratinocytes to a petroleum gauze [8, 11]. Because of their high cost and limited reliability, fragility, susceptibility to infection, and complexity of postoperative care, they are of limited use in large, full-thickness wounds that cover > 30% of the body surface [11, 53, 54]. Since it takes years for a dermal layer to form, Epicel™ is often used with a permanent dermal substitute. Nevertheless, it is often considered impractical for dermatological use [11].

5.2.2 EpiSkin™

EpiSkin™ (EpiSkin, Lyon, France) is an RHE model adopted by L'Oréal in 1997 and marketed as a 12-well plate system [46]. It consists of a bovine collagen matrix, human collagen and cultured human keratinocytes. Even though EpiSkin™ production is highly standardized, it was observed that inter-batch reproducibility was lower than intra-batch reproducibility [55]. The viable cells of the EpiSkin™ model are organized differently from the native epidermis [44, 56]. The two subtypes, irritation and penetration, differ in the thickness packing of the *stratum corneum*, with the penetration model having a denser packing and thicker characteristics. The phospholipid content of the EpiSkin™ irritation model is consistent with that of the human epidermis; however, this correlation of phospholipid content is lower in the penetration model. The number of free fatty acids (FFAs) and cholesterol esters in the models was lower than in native tissue; the concentration of surfactants required to elicit a response in the model was lower than that needed in vivo [56]. A wide variation in phospholipid content between batches was observed in the models [56, 57]. Although there is no direct correlation between any particular lipid class and drug transport, an increase in triglycerides and lipid retention within stratified cells was found to be associated with hyperproliferation and impaired barrier function [46, 58]. In the presence of vitamin C and lipids, better human skin models can be grown. Vitamin C plays a critical role in forming barrier lipids of the *stratum corneum* in human skin models, as evidenced by the improved barrier properties of Episkin™ [59, 60]. In permeability studies using caffeine as a reference solution, EpiSkin™ proved to be a suitable human skin model providing highly reproducible penetration data in good agreement with the penetration properties of native human skin [61].

5.2.3 SkinEthic™

SkinEthic™ (Martin Rosdy Laboratories, Nice, France) is an artificial human skin model consisting of a *stratum corneum, stratum granulosum* and *stratum spinosum*. Electron microscopic visualization of the model reveals the presence of components of the *lamina densa*, formerly called basal lamina, and anchoring filaments that mimic the actual basement membrane [44, 46]. Although the overall lipid composition of SkinEthic™ is consistent with that of native tissue, it differs from native human skin in the distribution of lipid droplets [57]. To predict chemical-induced skin irritancy and toxicity, the SkinEthic™ reconstituted epidermis model predicted well the phototoxic potential of chemicals under non-optimized conditions. The findings of toxicity studies have also revealed that the SkinEthic™ can be irradiated with high doses of UVA without detectable loss in tissue viability. This "insensitivity" on UVA irradiation can be ascribed to a fully differentiated *stratum corneum* in the SkinEthic™. This enabled a normal and functional permeability barrier for xenobiotics and light irradiation. In light of that, it has a much greater potential for in vitro phototoxicity prediction for all types of compounds, including low-phototoxicity compounds, when a higher dose of UV is applied [62]. Similarly, SkinEthic™ has also demonstrated its reliability in predicting the irritation potential of topical products. In addition, it is a useful model in a two-step strategy for screening acute and chronic irritation potential for the selection of vehicles for new topical drugs [63]. In percutaneous permeation and absorption studies, SkinEthic™ enabled the application and evaluation of raw materials contained in a formulation and their metabolic capacities due to a barrier function with a well-differentiated *stratum corneum*. The ranking of the topically applied compounds (mannitol, lauric acid and caffeine) in terms of permeation and skin absorption was comparable to ex vivo human skin. These results confirmed the testing potential of SkinEthic™ for in vitro permeation and percutaneous absorption evaluations of topical products [55].

5.2.4 EpiDerm™

EpiDerm™ (MatTek Corporation, Ashland, Maryland, USA) is described as a polycarbonate membrane seeded with normal human-derived epidermal keratinocytes (NHEK) to form a multilayered, highly differentiated model of the human epidermis [44, 46]. In terms of morphology, EpiDerm™ resembles the human epidermis, except that the keratinocytes do not have rete ridges as found in the native epidermal-dermal junction [44]. The lipid composition is comparable to native human skin but with lower cholesterol esters and FFAs [57]. In preliminary toxicity [56, 57] and permeation studies [55, 61], EpiDerm™ performed similarly to the other two marketed RHEs (EpiSkin™ and SkinEthic™). However, for permeation properties, prolonged hydrogel application to the skin model resulted in a decrease in permeation [61]. However, in studies of irritation, the simulation performance of EpiDerm™ is not entirely conclusive [64, 65]. For example, EpiDerm™ cultures deviated from the response of human skin, and it was concluded that this was due to the suboptimal barrier function of the model compared to excised skin. Topical application of an irritant was significantly toxic in EpiDerm™ cultures at much lower concentrations than in excised skin. Nevertheless, EpiDerm™ could serve as a predictive model for screening for irritation. However, great caution should be exercised in interpreting the results, as EpiDerm™ cultures do not have a competent barrier function and therefore require lower irritant concentrations than in in vivo or ex vivo studies to induce cytotoxic effects [64]. In another study, EpiDerm™ demonstrated usefulness for in vitro assessing the irritation potential of a series of cosmetic products. It allowed the measurement of quantifiable and objective endpoints relevant to irritant phenomena in vivo a [65].

5.2.5 *Dermagraft*®

Dermagraft® (Organogenesis, Canton, Maryland, USA) is a bioabsorbable, cryopreserved, human fibroblast-derived dermal substitute (see Fig. 5.1d). The dermal matrix is synthesized by culturing neonatal fibroblasts in a PLGA mesh. This mesh serves as a scaffold for producing cytokines, GFs, matrix proteins and collagen to form a 3D matrix [15, 66]. Apart from fibroblasts, Dermagraft® contains no other skin cells such as keratinocytes, endothelial cells, hair follicles or white blood cells. When placed in a wound, it stimulates fibrovascular growth and re-epithelialization [12]. Although it is FDA-approved as an interactive wound and burn dressing, it was classified as a high-risk class III product and requires clinical data to support its claims for use. It is recommended that this material should be used in patients who have an adequate blood supply [15]. It can be used as a temporary or permanent cover to support incorporating mesh split-thickness skin grafts on excised burn wounds [14, 15]. This cellular skin substitute seems to perform as well as allografts in terms of wound infection, wound exudate, wound healing time, wound closure, and graft uptake, and is easier to remove than allografts, with significantly higher patient satisfaction [67]. Advantages of this skin substitute include good tear resistance, ease of handling, and lack of rejection [68].

5.2.6 *TransCyte*™

TransCyte™ (Smith & Nephew, Watford, UK) is a bilaminar material consisting of a nylon mesh connected to a thin, semipermeable silicone membrane, with neonatal fibroblasts seeded onto the collagen-coated nylon mesh (see Fig. 5.1d). The fibroblasts are not viable upon application, and the nylon mesh is not biodegradable, so the material is designed as a temporary cover [51]. TransCyte™ for temporary coverage of partial thickness burns results in fewer dressing changes and less hypertrophic scarring than conventional treatment with topical silver sulfadiazine [15, 51, 69, 70].

5.2.7 Hyalograft3D™

Hyalograft3D™ (Fidia Advanced Biopolymers, Abano Terme, Italy) consists of Hyaff® (benzyl ester of HA) fibres processed into a 3D scaffold in which autologous fibroblasts find an ideal environment for adhesion, proliferation and subsequent production of a dermal ECM [71, 72]. Hyalograft 3D™ does not have a pseudo-epidermal layer (see Fig. 5.1c), but the product's action is enhanced by the cultured autologous fibroblasts that provide GFs and cytokines to the healing wound. It also deposits ECM components that 'condition' the wound for split skin grafting. It could also improve in vitro epithelial organization and maturation of the dermal–epidermal junction in organotypic skin bioconstructs [72]. Clinically, Hyalograft3D™ is primarily used to treat ulcers on the feet in combination with autologous epidermal bioconstructs, such as Laserskin® [71, 72]. Its use in the treatment of diabetic ulcers has been reported with significant improvement in wound closure compared to other devices [73].

5.2.8 Apligraf®

Apligraf® (Organogenesis, Canton, Maryland, USA) is a bilayer "living" skin equivalent (see Fig. 5.1f). It is composed of bovine type I collagen, allogeneic keratinocytes, and neonatal fibroblasts. As such, it has both allographic and xenographic characteristics. Since it contains both keratinocytes and fibroblasts, it allows cross-talk between the different cell types. In addition, this graft can produce its own matrix proteins and GFs [12, 14, 23]. During the production of Apligraf®, the epidermis is exposed to the air–liquid interface, allowing keratinocytes to stratify and form a stratum corneum. Apligraf® is indicated for partial- to full-thickness burns, skin grafts, chronic wounds, diabetic ulcers, and epidermolysis bullosa (a group of rare inherited skin diseases in which the skin becomes very fragile; any trauma or friction to the skin can cause painful blisters) [14, 74]. Although this product does not cause immunological rejection, the allogeneic cells of the construct do not survive in vivo for more than one to two months. Originally marketed as an organotypic skin substitute, Apligraf® is now considered a temporary bioactive dressing. It releases ECM components to the wound bed, together with cytokines and GFs. Some authors consider it an alternative to traditional skin grafting in partial-thickness burns. However, the product cannot be used for definitive wound closure in full-thickness injuries due to the transient nature of the grafted allogeneic cells. It, therefore, must be grafted along with an autologous epithelial source [72].

5.2.9   EpiDermFT™

EpiDermFT™ (MatTek Corporation, Ashland, Maryland, USA) features neonatal human dermal fibroblasts (NHFBs) and NHEKs in coculture, forming a multilayered epidermal-dermal model with well-differentiated *stratum corneum*, epidermis and dermis [44, 75, 76]. The dermal compartment consists of an EpiDermFT™ collagen matrix containing viable normal human dermal fibroblasts, and keratinocytes are cultured on the dermal component to form the epidermis. Ultra-structurally, this full-thickness skin model is very similar to human skin and thus provides a useful in vitro means to evaluate skin irritation and toxicity [76, 77]. The interactions between the epidermis and dermis would significantly affect the secretion of cytokines and chemokines after exposure to an irritant signal. These interactions are important for understanding skin irritation processes. There are significant differences in the secretion of pro-inflammatory mediators depending on whether dermal fibroblasts are present or not. Overall, the EpiDermFT™ model proved to be suitable for assessing skin irritation by chemicals. Furthermore, these models were sensitive to the damage caused by these chemicals and could be used to understand the structure–activity relationship (SAR) of different irritant chemicals. The in vitro culture results are consistent with previous in vivo studies conducted in a laboratory setting [76].

5.2.10   StrataTest®

The StrataTest® human skin model (Stratatech, Mallinckrodt Pharmaceuticals, Madison, Wisconsin, USA) was developed to simulate the architecture of human skin and its biological responses. This 3D full-thickness skin model contains dermal and epidermal compartments separated by an intact basement membrane. The presence of a fully stratified epidermis and a dermis populated with fibroblasts allows for cell-extracellular matrix interaction and epithelial-mesenchymal signalling. Paracrine signalling between keratinocytes and dermal fibroblasts can influence the expression of proteases, protease inhibitors, extracellular matrix components, cytoskeletal proteins, and numerous other cell signalling molecules. As such, it is essential for skin homeostasis [78]. It uses a near-diploid human keratinocyte cell line, NIKS® keratinocytes, a consistent and unlimited source of non-tumorigenic, pathogen-free p53 wild-type human keratinocyte progenitor cells [44, 78, 79]. This cell line can undergo normal epidermal differentiation and form a fully stratified epithelium similar to that of the natural human epidermis [78, 80]. Analysis of the StrataTest® human skin model confirmed tissue formation typical of the interfollicular epidermis. Furthermore, it showed the expression and corresponding localization of proteins essential for cell–cell adhesion, basement membrane formation, and epidermal function. The StrataTest® model has a variety of applications in toxicology testing. Independent batches of skin tissue responded consistently to known chemical irritants, including reactive oxygen species (ROS) formation after exposure to ozone, cigarette smoke and UV radiation [44, 78].

References

1. Yu JR, Navarro J, Coburn JC, Mahadik B, Molnar J, Holmes JH IV, Nam AJ, Fisher JP (2019) Current and future perspectives on skin tissue engineering: key features of biomedical research, translational assessment, and clinical application. Adv Healthcare Mater 8(5):1801471
2. Mansbridge JN (2009) Tissue-engineered skin substitutes in regenerative medicine. Curr Opin Biotechnol 20(5):563–567
3. Vig K, Chaudhari A, Tripathi S, Dixit S, Sahu R, Pillai S, Dennis VA, Singh SR (2017) Advances in skin regeneration using tissue engineering. Int J Mol Sci 18(4)
4. Lu KW, Khachemoune A (2022) Skin substitutes for the management of mohs micrographic surgery wounds: a systematic review. Arch Dermatol Res 1–15
5. Nilforoushzadeh MA, Amirkhani MA, Khodaverdi E, Razzaghi Z, Afzali H, Izadpanah S, Zare S (2022) Tissue engineering in dermatology-from lab to market. Tissue Cell 74:101717
6. Chen C, He J, Huang J, Yang X, Liu L, Wang S, Ji S, Chu B, Liu W (2022) Fetal dermis inspired parallel PCL fibers layered PCL/COL/HA scaffold for dermal regeneration. Reactive Funct Polym 170:105146
7. Rajalekshmy G, Rekha M (2021) Trends in bioactive biomaterials in tissue engineering and regenerative medicine. Biomater Tissue Eng Regenerative Med, Springer 2021, 271–303
8. Murray RZ, West ZE, Cowin AJ, Farrugia BL (2019) Development and use of biomaterials as wound healing therapies. Burns trauma 7
9. Bello YM, Falabella AF, Eaglstein WH (2001) Tissue-engineered skin. Am J Clin Dermatol 2(5):305–313
10. Debels H, Hamdi M, Abberton K, Morrison W (2015) Dermal matrices and bioengineered skin substitutes: a critical review of current options. Plastic Reconstruct Surgery Global Open 3(1)
11. Dai C, Shih S, Khachemoune A (2020) Skin substitutes for acute and chronic wound healing : an updated review. J Dermatol Treat 31(6):639–648
12. Cahn B, Lev-Tov H (2020) Cellular-and acellular-based therapies: skin substitutes and matrices, local wound care for dermatologists. In: Alavi A, Maibach H (eds) Updates in Clinical Dermatology, 139–151
13. Nicholas MN, Jeschke MG, Amini-Nik S (2016) Methodologies in creating skin substitutes. Cell Mol Life Sci 73(18):3453–3472
14. Shukla A, Dey N, Nandi P, Ranjan M (2015) Acellular dermis as a dermal matrix of tissue engineered skin substitute for burns treatment. Ann Public Health Res 2(3):1023
15. Urciuolo F, Casale C, Imparato G, Netti PA (2019) Bioengineered skin substitutes: the role of extracellular matrix and vascularization in the healing of deep wounds. J Clin Med 8(12):2083
16. Van der Veen VC, van der Wal MB, van Leeuwen MC, Ulrich MM, Middelkoop E (2010) Biological background of dermal substitutes. Burns 36(3):305–321
17. Catalano E, Cochis A, Varoni E, Rimondini L, Azzimonti B (2013) Tissue-engineered skin substitutes: an overview. J Artif Organs 16(4):397–403
18. Wainwright D (1995) Use of an acellular allograft dermal matrix (AlloDerm) in the management of full-thickness burns. Burns 21(4):243–248
19. Deneve JL, Turaga KK, Marzban SS, Puleo CA, Sarnaik AA, Gonzalez RJ, Sondak VK, Zager JS (2013) Single-institution outcome experience using AlloDerm® as temporary coverage or definitive reconstruction for cutaneous and soft tissue malignancy defects. Am Surg 79(5):476–482
20. Gordley K, Cole P, Hicks J, Hollier L (2009) A comparative, long term assessment of soft tissue substitutes: AlloDerm, Enduragen, and Dermamatrix. J Plast Reconstr Aesthet Surg 62(6):849–850

21. Breuing KH, Warren SM (2005) Immediate bilateral breast reconstruction with implants and inferolateral AlloDerm slings. Ann Plast Surg 55(3):232–239
22. Buinewicz B, Rosen B (2004) Acellular cadaveric dermis (AlloDerm): a new alternative for abdominal hernia repair. Ann Plast Surg 52(2):188–194
23. Oualla-Bachiri W, Fernández-González A, Quiñones-Vico MI, Arias-Santiago S (2020) From grafts to human bioengineered vascularized skin substitutes. Int J Mol Sci 21(21):8197
24. Taufique ZM, Bhatt N, Zagzag D, Lebowitz RA, Lieberman SM (2019) Revascularization of alloderm used during endoscopic skull base surgery. J Neurol Surg Part B: Skull Base 80(01):046–050
25. Becker S, Saint-Cyr M, Wong C, Dauwe P, Nagarkar P, Thornton JF, Peng Y (2009) Allo-Derm versus DermaMatrix in immediate expander-based breast reconstruction: a preliminary comparison of complication profiles and material compliance. Plast Reconstr Surg 123(1):1–6
26. Troy J, Karlnoski R, Downes K, Brown KS, Cruse CW, Smith DJ, Payne WG (2013) The use of EZ Derm® in partial-thickness burns: an institutional review of 157 patients. Eplasty 13
27. Bano F, Barrington J, Dyer R (2005) Comparison between porcine dermal implant (Permacol) and silicone injection (Macroplastique) for urodynamic stress incontinence. Int Urogynecol J 16(2):147–150
28. MacLeod T, Cambrey A, Williams G, Sanders R, Green C (2008) Evaluation of Permacol™ as a cultured skin equivalent. Burns 34(8):1169–1175
29. Mahadavan L, Veeramootoo D, Daniels I (2008) Early results of porcine collagen membrane (Permacol) in pelvic floor reconstruction following prone cyclindrical PAE. Eur J Surg Oncol 34(10):1165
30. Pitkin L, Rimmer J, Lo S, Hosni A (2008) Aesthetic augmentation rhinoplasty with permacol: how we do it. Clin Otolaryngol 33(6):615–618
31. Healy C, Boorman J (1989) Comparison of EZ Derm and Jelonet dressings for partial skin thickness burns. Burns 15(1):52–54
32. Hodde J, Ernst D, Hiles M (2005) An investigation of the long-term bioactivity of endogenous growth factor in OASIS wound matrix. J Wound Care 14(1):23–25
33. Mostow EN, Haraway GD, Dalsing M, Hodde JP, King D, Group OVUS (2005) Effectiveness of an extracellular matrix graft (OASIS Wound Matrix) in the treatment of chronic leg ulcers: a randomized clinical trial. J Vascular Surg 41(5):837–843
34. Lagus H, Sarlomo-Rikala M, Böhling T, Vuola J (2013) Prospective study on burns treated with Integra®, a cellulose sponge and split thickness skin graft: comparative clinical and histological study—randomized controlled trial. Burns 39(8):1577–1587
35. Moiemen NS, Vlachou E, Staiano JJ, Thawy Y, Frame JD (2006) Reconstructive surgery with Integra dermal regeneration template: histologic study, clinical evaluation, and current practice. Plast Reconstr Surg 117(7S):160S-174S
36. Jackson SR, Roman S (2019) Matriderm and split skin grafting for full-thickness pediatric facial burns. J Burn Care Res 40(2):251–254
37. Boyce ST, Lalley AL (2018) Tissue engineering of skin and regenerative medicine for wound care. Burns Trauma 6
38. Kamel RA, Ong JF, Eriksson E, Junker JP, Caterson EJ (2013) Tissue engineering of skin. J Am Coll Surg 217(3):533–555
39. Still J Jr, Orlet H, Law E (1994) Use of cultured epidermal autografts in the treatment of large burns. Burns 20(6):539–541
40. Ronfard V, Rives J-M, Neveux Y, Carsin H, Barrandon Y (2000) Long-term regeneration of human epidermis on third degree burns transplanted with autologous cultured epithelium grown on a fibrin matrix1, 2. Transplantation 70(11):1588–1598
41. Danilenko DM, Phillips GDL, Diaz D (2016) In vitro skin models and their predictability in defining normal and disease biology, pharmacology, and toxicity. Toxicol Pathol 44(4):555–563
42. Setijanti HB, Rusmawati E, Fitria R, Erlina T, Adriany R (2019) Development the technique for the preparation and characterization of reconstructed human epidermis (RHE). Springer, Alternatives to Animal Testing, pp 20–32

43. Rehder J, Souto LRM, Issa CMBM, Puzzi MB (2004) Model of human epidermis reconstructed in vitro with keratinocytes and melanocytes on dead de-epidermized human dermis. Sao Paulo Med J 122:22–25

44. Suhail S, Sardashti N, Jaiswal D, Rudraiah S, Misra M, Kumbar SG (2019) Engineered skin tissue equivalents for product evaluation and therapeutic applications. Biotechnol J 14(7):1900022

45. Martínez-Santamaría L, Guerrero-Aspizua S, Del Río M (2012) Skin bioengineering: preclinical and clinical applications. Actas Dermo-Sifiliográficas (English Edition) 103(1):5–11

46. Netzlaff F, Kaca M, Bock U, Haltner-Ukomadu E, Meiers P, Lehr C-M, Schaefer UF (2007) Permeability of the reconstructed human epidermis model Episkin® in comparison to various human skin preparations. Eur J Pharm Biopharm 66(1):127–134

47. Gushiken LFS, Beserra FP, Bastos JK, Jackson CJ, Pellizzon CH (2021) Cutaneous wound healing: an update from physiopathology to current therapies. Life 11(7):665

48. Kaur A, Midha S, Giri S, Mohanty S (2019) Functional skin grafts: where biomaterials meet stem cells. Stem Cells Int 2019

49. Goodarzi P, Falahzadeh K, Nematizadeh M, Farazandeh P, Payab M, Larijani B, Beik AT, Arjmand B (2018) Tissue engineered skin substitutes. Cell Biol Transl Med 3:143–188

50. Nicholas MN, Yeung J (2017) Current status and future of skin substitutes for chronic wound healing. J Cutan Med Surg 21(1):23–30

51. Thornton JF, Gosman A (2004) Skin grafts and skin substitutes. Sel Readings Plastic Surg 10(1):1–24

52. Guenou H, Nissan X, Larcher F, Feteira J, Lemaitre G, Saidani M, Del Rio M, Barrault CC, Bernard F-X, Peschanski M (2009) Human embryonic stem-cell derivatives for full reconstruction of the pluristratified epidermis: a preclinical study. The Lancet 374(9703):1745–1753

53. Límová M (2010) Active wound coverings: bioengineered skin and dermal substitutes. Surg Clin 90(6):1237–1255

54. Nathoo R, Howe N, Cohen G (2014) Skin substitutes: an overview of the key players in wound management. J Clin Aesthetic Dermatol 7(10):44

55. Lotte C, Patouillet C, Zanini M, Messager A, Roguet R (2002) Permeation and skin absorption: reproducibility of various industrial reconstructed human skin models. Skin Pharmacol Physiol 15(Suppl. 1):18–30

56. Ponec M, Boelsma E, Weerheim A, Mulder A, Bouwstra J, Mommaas M (2000) Lipid and ultrastructural characterization of reconstructed skin models. Int J Pharm 203(1–2):211–225

57. Ponec M, Boelsma E, Gibbs S, Mommaas M (2002) Characterization of reconstructed skin models. Skin Pharmacol Physiol 15(Suppl. 1):4–17

58. Schäfer-Korting M, Bock U, Gamer A, Haberland A, Haltner-Ukomadu E, Kaca M, Kamp H, Kietzmann M, Korting HC, Krächter H-U (2006) Reconstructed human epidermis for skin absorption testing: results of the German prevalidation study. Altern Lab Anim 34(3):283–294

59. Ponec M, Weerheim A, Kempenaar J, Mulder A, Gooris GS, Bouwstra J, Mommaas AM (1997) The formation of competent barrier lipids in reconstructed human epidermis requires the presence of vitamin C. J Invest Dermatol 109(3):348–355

60. Auger FA, Pouliot R, Tremblay N, Guignard R, Noël P, Juhasz J, Germain L, Goulet F (2000) Multistep production of bioengineered skin substitutes: sequential modulation of culture conditions. In vitro Cellular Develop Biol-Animal 36(2):96–103

61. Dreher F, Fouchard F, Patouillet C, Andrian M, Simonnet J-T, Benech-Kieffer F (2002) Comparison of cutaneous bioavailability of cosmetic preparations containing caffeine or α-tocopherol applied on human skin models or human skin ex vivo at finite doses. Skin Pharmacol Physiol 15(Suppl. 1):40–58

62. Bernard F, Barrault C, Deguercy A, De Wever B, Rosdy M (2000) Development of a highly sensitive in vitro phototoxicity assay using the SkinEthicTM reconstructed human epidermis. Cell Biol Toxicol 16(6):391–400

63. de Fraissinette AdB, Picarles V, Chibout S, Kolopp M, Medina J, Burtin P, Ebelin M, Osborne S, Mayer F, Spake A (1999) Predictivity of an in vitro model for acute and chronic skin irritation (SkinEthic) applied to the testing of topical vehicles. Cell Biol Toxicol 15(2):121–135

64. Gibbs S, Vietsch H, Meier U, Ponec M (2002) Effect of skin barrier competence on SLS and water-induced IL-1α expression. Exp Dermatol 11(3):217–223
65. Faller C, Bracher M, Dami N, Roguet R (2002) Predictive ability of reconstructed human epidermis equivalents for the assessment of skin irritation of cosmetics. Toxicol In Vitro 16(5):557–572
66. Marston WA, Hanft J, Norwood P, Pollak R (2003) The efficacy and safety of Dermagraft in improving the healing of chronic diabetic foot ulcers: results of a prospective randomized trial. Diabetes Care 26(6):1701–1705
67. Hansbrough JF, Mozingo DW, Kealey GP, Davis M, Gidner A, Gentzkow GD (1997) Clinical trials of a biosynthetic temporary skin replacement, dermagraft-transitional covering, compared with cryopreserved human cadaver skin for temporary coverage of excised burn wounds. J Burn Care Rehabil 18(1):43–51
68. Hart CE, Loewen-Rodriguez A, Lessem J (2012) Dermagraft: use in the treatment of chronic wounds. Adv Wound Care 1(3):138–141
69. Kumar RJ, Kimble RM, Boots R, Pegg SP (2004) Treatment of partial-thickness burns: a prospective, randomized trial using TranscyteTM. ANZ J Surg 74(8):622–626
70. Noordenbos J, Doré C, Hansbrough JF (1999) Safety and efficacy of TransCyte* for the treatment of partial-thickness burns. J Burn Care Rehabil 20(4):275–281
71. Giuggioli D, Sebastiani M, Cazzato M, Piaggesi A, Abatangelo G, Ferri C (2003) Autologous skin grafting in the treatment of severe scleroderma cutaneous ulcers: a case report. Rheumatology 42(5):694–696
72. Shevchenko RV, James SL, James SE (2010) A review of tissue-engineered skin bioconstructs available for skin reconstruction. J R Soc Interface 7(43):229–258
73. Santema TK, Poyck PP, Ubbink DT (2016) Systematic review and meta-analysis of skin substitutes in the treatment of diabetic foot ulcers: highlights of a Cochrane systematic review. Wound Repair and Regeneration 24(4):737–744
74. Chan EC, Kuo S-M, Kong AM, Morrison WA, Dusting GJ, Mitchell GM, Lim SY, Liu G-S (2016) Three dimensional collagen scaffold promotes intrinsic vascularisation for tissue engineering applications. PLoS ONE 11(2):e0149799
75. Semlin L, Schäfer-Korting M, Borelli C, Korting HC (2011) In vitro models for human skin disease. Drug Discovery Today 16(3–4):132–139
76. Mallampati R, Patlolla RR, Agarwal S, Babu RJ, Hayden P, Klausner M, Singh MS (2010) Evaluation of EpiDerm full thickness-300 (EFT-300) as an in vitro model for skin irritation: studies on aliphatic hydrocarbons. Toxicol In Vitro 24(2):669–676
77. Hayden PJ, Petrali JP, Stolper G, Hamilton TA, Jackson GR Jr, Wertz PW, Ito S, Smith WJ, Klausner M (2009) Microvesicating effects of sulfur mustard on an *in vitro* human skin model. Toxicol In Vitro 23(7):1396–1405
78. Rasmussen C, Gratz K, Liebel F, Southall M, Garay M, Bhattacharyya S, Simon N, Vander Zanden M, Van Winkle K, Pirnstill J (2010) The StrataTest® human skin model, a consistent *in vitro* alternative for toxicological testing. Toxicol in vitro 24(7):2021–2029
79. Zhang Z, Michniak-Kohn BB (2012) Tissue engineered human skin equivalents. Pharmaceutics 4(1):26–41
80. Allen-Hoffmann BL, Schlosser SJ, Ivarie CA, Meisner LF, O'Connor SL, Sattler CA (2000) Normal growth and differentiation in a spontaneously immortalized near-diploid human keratinocyte cell line NIKS. J Investigative Dermatol 114(3):444–455

Chapter 6
Applications

3D skin equivalents were one of the first TE organs used for clinical trials. In early skin models, fibroblasts and keratinocytes were incorporated into a nylon mesh to mimic the dermal matrix [1–3]. Since then, the evolution of scaffolds for skin cell culture has spurred the development of more sustainable and biologically accurate biomaterials that allow for the mimicking of properties and prediction of skin behaviour [2]. Despite advances in this field, there is no consensus on how well artificial skin can mimic the behaviour of natural skin [4–6]. In addition, many commercially available skin substitutes have been reported to lack mechanical and textural similarity because they require replicating the biological and histological properties of skin [7]. Nevertheless, these efforts have resulted in sophisticated in vitro skin models widely used for clinical applications, advances in wound healing, and as test systems for pharmaceutical and cosmetic research [1, 2, 8–13].

6.1 Skin Reconstruction

In general, all approaches to wound healing aim to create an environment that minimizes infection, promotes proper moisture balance and facilitates re-epithelialization of the wound [14]. The main application of bioengineered skin models is to augment or replace skin grafts used to treat patients with severe skin injuries. In the clinical setting, skin grafts can be used to treat extensive tissue defects by restoring normal barrier function while stimulating wound healing responses [10, 12, 15]. Currently, the clinical "gold standard" in treating full-thickness skin injuries remains autologous split-thickness skin grafting. In this procedure, the epidermis with a superficial portion of the dermis is harvested from an undamaged skin donor site and applied to the full-thickness wound. When applied to the wound, the capillaries of the split skin graft (SSG) form anastomoses. In other words, they connect to the existing capillary network to supply nutrients to the graft. This is called "graft uptake" [16, 17]. The donor site heals similarly to the superficial partial thickness wound by migration of keratinocytes from hair follicles, sweat glands, and wound edges [16]. However, if normal tissue healing is impaired or sufficient healthy donor tissue is unavailable, TE constructs may be required [10, 12, 15]. Due to the great importance and demand for skin replacement products, there is a long history of material development with an emphasis on creating biomaterials for skin substitution. This is because it has still not been possible to produce a suitable TE substrate that can satisfactorily replicate the epidermal and dermal niches in vivo to fulfil aesthetic and functional requirements. TE has seen exponential growth in recent years, with several "off-the-shelf" dermal and epidermal substitutes now available. While some products have been shown to reduce morbidity and improve clinical outcomes following injury, no single skin substitute currently on the market has been shown to fully restore normal skin structure and physiological function of native uninjured skin [12, 16, 18, 19].

When the skin is extensively injured, it loses its ability to prevent bacterial infection and regulate temperature or fluid transport [12, 16, 18, 20]. The natural response to severe skin injury in adults, which involves granulation and re-epithelialization of the tissue, is characterized by the rapid proliferation of fibroblasts that deposit randomly oriented collagen fibres to fill the tissue defect. This is followed by the migration of keratinocytes and contraction of myofibroblasts that restore the barrier [21]. This accumulation of disorganized tissue results in the formation of a fibrotic scar, often associated with a lack of sensation and elasticity, as well as defective features. As it turns out, the "healing" does not restore the native skin function, histologic structure, or esthetics [12, 16, 18, 20, 22].

Another point is that other cell types normally present in the skin may slower regenerate or may not regrow at all. For example, even when sebaceous glands are transplanted into skin grafts, normal secretory function usually does not return for months [12, 14, 15]. Similarly, sensory and autonomic nerves present in adjacent areas of healthy skin may grow in and eventually re-nerve the wound area, but this process is slow and never complete [15]. As a result, these "healed" skin areas may experience abnormal sensation or sweat function. Finally, and probably more importantfrom the patient's perspective, the loss of melanocytes leads to changes in skin pigmentation that can be disfiguring and difficult to treat with current cosmetic techniques [23]. Advanced therapies to combat acute and chronic skin wounds are likely to come about through our knowledge of regenerative medicine combined with appropriate TE skin substitutes [24].

6.1.1 *"Bioactive" Guides for Skin Regeneration*

Ideally, a successful TE skin construct replicates the complexity of the natural 3D structure and fulfils the functions of the natural skin tissue. In addition, it should support vascularization and provide supportive cues to cells in the local environment. Finally, when implanted in vivo, it must also integrate into the host with minimal scarring while eliciting a controlled inflammatory response [12, 25].

The role of engineered skin substitutes in treating burns and chronic wounds continues to evolve. New products are continuously developed, manufactured and approved for clinical use [26]. Early approaches used synthetic components to minimize fluid loss and mechanical stress while maintaining structural stability at the wound site. Nylon and silicone composites proved popular and could be coated with biomolecules and skin cells, leading to the creation of products such as Biobrane®, TransCyte™ and Integra®. The advantage of these materials over traditional wound dressings is that the collagen forms a hydrogel that protects the wound while the scaffold absorbs fluid [14, 19, 26, 27].

Scaffolds composed entirely of natural materials have also gained popularity. They contain protein motifs such as the arginine-glycine-aspartic acid (RGD) and alanine-glycine-aspartic acid-valine (AGDV) sequences that facilitate cell adhesion and exhibit better compatibility and degradation in vivo. This is especially true when they contain biomolecules already naturally part of the skin ECM [14]. These constructs can be realised by incorporating GFs and cells of interest (usually fibroblasts, keratinocytes, or stem cells grown in vitro) to facilitate ingrowth of native cells or proliferation of seeded cells from autologous or allogeneic sources. Such GF-loaded or cell-loaded hydrogels are often used to study skin properties such as immunoreactivity [28], wound closure [29], epithelialization [30, 31], angiogenesis [31, 32], or hair growth [32, 33]. ECM-based scaffolds are commonly used in vitro to model aspects of skin physiology and transport phenomena to exploit the characteristic properties of protein-based materials. While these models are useful for studying specific properties of native skin tissue, they are generally not comprehensive because they take a narrow approach to a single target and neglect the complexity of skin physiology as a whole. For example, Sakamoto et al. [31] used a pliable gelatin hydrogel sheet that maintained the release of bFGF and conformed to the shape of the wound. This construct was shown to accelerate epithelialization, granulation tissue formation, and angiogenesis in mice [31]. This and other similar 3D skin models need further development to investigate specific features such as stratified or vascularized tissue formation before they can be successfully translated for clinical wound healing and tissue regeneration [12]. Compartmentalised as these models may be, they have justified using protein-based scaffolds in clinical trials that generally reflect the positive trends observed in vitro. In 2016, a cohort clinical study was conducted on the use of keratin-based scaffolds in superficial and partial thickness burn injuries. This clinical study was supported by preclinical animal studies and a clinical randomised control trial of partial thickness donor site wound healing and scar treatment. The keratin in this product line has been shown to stimulate keratinocyte activity by increasing migration and proliferation rates and upregulating the expression of key basement membrane proteins. Compared to the current clinical standard of care, keratin-based products have resulted in faster rates of re-epithelialization, reduced scarring and improved clinical parameters, such as shorter healing time, shorter inpatient treatment, shorter out-patient appointments, and reduced antibiotic use [34]. While many other commercial clinical ECM constructs (Dermagraft®, Integra®, and AlloDerm®) have been marketed as dermal equivalents or degradable dressings that help accelerate wound closure, cosmetic outcomes are still generally poor [35].

The transfer of a number of cellular functions to the macroscopic processes of scar formation and wound contraction is often difficult to achieve [12].

Engineered bilayered scaffolds, such as Apligraf® and MatriDerm®, typically recapitulate only the epidermis and dermis, making them ideal for addressing injuries such as first- and second-degree burns. As such, third- (involving the epidermis, dermis, and hypodermis) and fourth-degree burns (affecting all the layers down to the muscle and bone) are less often considered [12].

Although not currently used for clinical applications, several three-layer constructs with a hypodermis-like layer have been developed for in vitro models of human skin [36, 37]. These constructs are usually developed by culturing at the air–liquid interface and are used to study skin tissue properties such as barrier function or cell behaviour. The fabrication and regeneration of the hypodermis have not been fully explored and therefore holds great potential for both laboratory and clinical skin growth and regeneration. Previous studies have reported that incorporating other cell types, such as MSCs, into scaffolds can support these ventures [12].

6.1.2   The Regulatory Hurdles

The timeline for approval of innovative solutions can be quite long, as the translation from the laboratory to the bedside often requires regulatory approval to determine the safety and efficacy of the therapy, as well as a review of quality controls [12]. There are several regulatory hurdles to consider when translating tissue engineering scaffolds into the clinic, depending on the type of technology involved. An acellular scaffold should be subject to less regulatory scrutiny compared to approaches that use allogeneic or xenogeneic cells, iPS cells, embryonic stem cells, or even significant ex vivo manipulation of autologous cells. The introduction of cells as a component of TE carries risks related to potential immunogenicity, teratoma formation, cell culture adaptation/morphogenesis or contamination that need to be considered to ensure safety [35].

The scaffold material may influence the process of registering a new TE product. The development of highly bioactive scaffold materials with poorly defined degradation products may require a significant effort to demonstrate its safety and degradation products, increasing the time and cost of preclinical and clinical investigations. This can create a dichotomy, as attempts to establish novel systems that interact with biological systems in fundamentally new ways are often distinct from rapidly implementing technologies. The former can enable groundbreaking academic research, but the cost and effort required to implement fundamentally new technologies often can be a significant barrier. Ideally, the development of novel TE materials will lead to innovative functionality that cannot be replicated with existing materials. Then these transformative technologies can be advanced based on their ability to perform a specific function [35, 38].

A problem that arises, particularly with TE, is the extent to which preclinical studies, including theoretical modelling, in vitro characterization, and in vivo (animal) studies, are predictive of subsequent function and performance in humans [35]. Among important considerations are, for example, the interspecies variability of the immune system. These raise the concern that studies in rodents may not predict immune response to an implanted construct in humans because there are many differences between mouse and human immunology [39, 40]. There is evidence that the sex of the chosen laboratory animal may also influence its demonstrated efficacy [41]. Therefore, attempts have been made to produce humanized mice with immune systems more similar to humans, improving the predictive power of such models [42].

In the regulatory approval process by the US FDA, a new "therapeutic" product is traditionally approved through one of four different routes: for tissues, biologics, drugs, or medical devices. TE approaches present a particular challenge at this stage. Many TE strategies can fall into two, three, or even all four of these categories, depending on their sophistication. A simple scaffold used for a tissue filler could be classified as a pure medical device. However, once additional scaffold complexity is incorporated, the construct can no longer be classified as a pure medical device. Indeed, in most TE approaches, cells are seeded on a scaffold and may incorporate additional signalling elements, chemical functionalities, or the release of small molecules or proteins from the scaffold. From the outset, the precise regulatory pathway of a TE therapy is less clear. Often, aspects of the regulatory processes for each of these pathways must be considered to develop a single therapy. This can significantly increase the time and cost of regulatory approval. Currently, preclinical and early clinical technologies assessments are conducted on a case-by-case basis. This frequently requires innovation in the regulatory framework by companies and regulators to ultimately determine the best course of action to ensure safety [43, 44].

Regulatory approval processes are required for many products from TE, including skin products; either as Premarket Approval (PMA), Investigational New Drug (IND) or Biologics License Application (BLA) [12]. To help product developers navigate the regulatory system, the FDA's Center for Biologics Evaluation and Research (CBER) provides an overview of what is required to review their products, including preclinical trial design, evaluation results, and clinical trial phase flow [45, 46].

6.2 Modeling of Physiological Processes

In vitro skin substitutes use 3D arranged human cells to mimic cell–cell and cell–matrix interactions. Most models developed aim to model healthy skin with intact barrier properties, and only a few models mimic damaged skin [9]. Modelling normal skin conditions is used to better understand and predict in vivo interactions and study molecules that can enhance and facilitate them. Optimizing the wound healing process, preventing dangerous effects of UV radiation, and delaying the ageing process are physiological conditions that are frequently studied to gain better insight into the skin's unique properties and ultimately achieve a healthier life status [1, 11].

6.2.1 Epithelialization and Barrier Function

The epidermis consists of keratinized cells embedded in a stratum corneum lipid matrix, which protects against transcutaneous water loss and other physical, chemical, and biological irritants. Normalizing skin equivalent barrier formation is complex as many external factors differ between in vivo and in vitro culture conditions. In addition to nutrient imbalance, there are several differences in external factors. Some studies have reported altered barrier function in in vitro developed skin equivalents compared to native skin [1, 47]. To counteract this, Mieremet and his group [48] cultured skin equivalents in the presence of FFAs. By supplementing the culture media of human skin equivalents with palmitic acid (PA), a saturated FFA, they aimed to shed light on the role of PA supplementation on epidermal morphogenesis and barrier formation of in vitro skin models and to gain more insight into the presence of FFAs in their *stratum corneum*. In the presence of PA, enhanced epidermal morphogenesis was observed as it facilitated the proper execution of the early differentiation process in the epidermis. However, the lipid barrier composition remained the same regardless of the PA concentration and was still lower than in native skin [48]. It has been reported that subjecting human skin equivalents to hypoxia conditions reduces epidermal thickness. It further alters the lipid composition of the *stratum corneum* to more closely resemble natural human skin [47]. The tight junctions within the epidermal and dermal layers maintain skin barrier homeostasis. They act as a paracellular barrier beneath the *stratum corneum*, but little is known about the mechanisms behind this. Several studies have attempted to clarify the relationship between epidermal tight junctions and the cutaneous permeability barrier using skin equivalents treated with various inflammatory cytokines such as IL-4, I-17 and IL-22 [49, 50]. For example, disrupted tight junctions have been associated with the impaired formation of the *stratum corneum*, which is responsible for the skin barrier. This suggests the possibility that disrupted tight junctions affect polar lipids and profilaggrin processing by disturbing the pH of the *stratum corneum* [49]. In addition, some factors related to the pathogenesis of atopic dermatitis (AD) induce tight junction dysfunction. It was found that only IL-17 impairs the tight junction barrier, resulting in a defect in the degradation of filaggrin monomers in the IL-17- treated skin model. It seems that the tight junctions at AD are at least partially disrupted by the action of IL-17, which may result in an abnormal barrier of the *stratum corneum* [50]. Also, IL-31 treatment of 3D human skin models resulted in a disrupted skin barrier phenotype resembling AD. This is due to IL-31 interference with keratinocyte differentiation and suppressed expression of markers of terminal differentiation. On the other hand, the structure and function of the physical skin barrier in the disrupted skin barrier models recovered after daily topical treatment with a ceramide-containing ointment [51].

Repeatedly, full-thickness skin models have been used to create models of skin wounds or to characterize treatment procedures by their use. After complete tissue dissection, re-epithelialization at the wound edges occurred within 8–12 h, while it took four days for the epidermis to be fully restored [13]. Rapid restoration of the epithelial barrier prevents excessive water loss and bacterial infections, reducing mortality in patients who have lost a significant amount of total body surface area. To investigate the molecular mechanisms of 3D cell migration associated with wound healing and re-epithelialization, Geer and colleagues [52] prepared epidermal equivalents that contained fibrin as a substrate for keratinocytes. This model system mimicked several aspects of the re-epithelialization process in vivo. These included the length of the lag phase of keratinocyte activation, the time of healing, the spatiotemporal pattern of cell proliferation in and around the wound, the stage of cell differentiation in the wound, and the reformation of the basement membrane. In addition, the authors have shown that fibrin gels of physiological composition promote the re-epithelialization of cut wounds in composite skin equivalents. Their findings suggested that such skin models could serve as a realistic toxicology model to optimize new therapeutic formulations for wound healing before animal testing [52]. To model the physiological response of cutaneous wound healing and UV radiation, Garcia and her group [53] developed a skin-humanized mouse model system (Fig. 6.1) consisting of bioengineered human skin implanted into immunodeficient mice.

Starting from skin cells (keratinocytes and fibroblasts) isolated from normal donor skin or patient biopsies, they have succeeded in deconstructing and reconstructing various inherited skin disorders, including genodermatoses. This is a hereditary disease of the skin, which includes various conditions that can affect the color, texture,

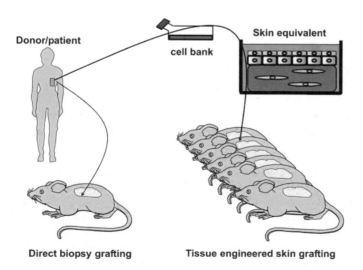

Fig. 6.1 Schematic diagram of the process used to produce skin tissue. Skin cells (fibroblasts and keratinocytes) are obtained from a donor/patient and placed in primary culture. Primary human keratinocytes are cultured on dermal fibrin equivalents (in which fibroblasts are embedded) to produce the skin equivalent. Portions of the skin equivalent can be transplanted into mice. In contrast, transplantation of an entire skin biopsy (direct grafting) would produce few grafted mice. Reproduced from [53] with permission from John Wiley & Sons, Inc.

and structural integrity of the epidermis, epidermal appendages, and connective tissue [54]) cancer-prone diseases in a large number of skin-humanized mice. Besides, the model enables studies to be carried out on normal human skin to gain further insights into physiological processes such as wound healing or UV responses [53].

Well-defined induced wounds served as an initial characterization of topically applied pharmaceutical agents (morphine) and drug delivery systems [55]. In experimental therapies, opioids were used for severe local pain, and nanocarriers, especially the solid lipid nanoparticle type, are options for slow drug release and improved skin penetration [56–58]. Experimental wounding allowed the demonstration of enhanced keratinocyte migration, as previously observed in keratinocyte monocultures [13].

6.2.2 Skin Ageing

Several mechanisms are involved in ageing. These include the accumulation of mutations in genetic material, the accumulation of toxic metabolites, the generation of free radicals that cause oxidative damage, chemical changes, and the crosslinking of macromolecules by glycation. Skin equivalents have also shown how senescence affects skin morphology and functionality. With age, visible histological changes in the skin occur. These include reduction in epidermis thickness, changes in dermis composition, disorganization of ECM components, and flattening of the dermo-epidermal junction [1, 10, 59].

Various strategies can be used to design in vitro skin ageing models. The most common forms are:

(i)   photoaging generation by UV light, where collagen is previously exposed to UV radiation to induce changes in the skin,
(ii) generation of chronological ageing by glycation or alteration of the collagen composition of the dermis, where the collagen is pre-incubated with sugars (ribose or glucose) before being used to prepare the reconstructed skin [60].

Understanding the contribution of the dermis to skin ageing is a key issue, as this tissue is particularly important to the integrity of the skin, and its properties can affect the epidermis [59, 61, 62]. Changes in fibroblast properties associated with ageing or senescence have been studied either in long-term cell cultures or using fibroblasts from skin biopsies obtained from donors with advancing age. The reported alterations in fibroblast properties observed during serial in vitro expansion may all participate to a greater or lesser extent in the age-related skin changes in vivo [59]. The complex network of intercellular crosstalk between different cell types in human skin is not fully understood, and it remains unclear how aged cells may adversely affect tissue function. These effects cannot simply be studied in a classical 2D system but require a more sophisticated 3D approach [63]. Thus, considering the different populations of dermal fibroblasts and their age-related changes, novel methods are developed to prepare in vitro skin models containing a dermal compartment populated with either papillary or reticular fibroblasts [59, 61–63]. In particular, populations of papillary fibroblasts seem to disappear or at least decrease with aging, whereas reticular fibroblasts do not change [59]. Moreover, epidermal differentiation and maturation were favoured and found to be optimal in the presence of papillary fibroblasts, which had little effect on ECM, in contrast to reticular fibroblasts, which had a significant positive effect on the production of ECM molecules of the dermal–epidermal junction and dermis. These functional differences could affect wound healing and skin ageing, considering that papillary fibroblasts in human skin slowly decrease with age. All in all, a reconstructed skin model with a dermis containing only reticular fibroblasts is a reasonable approach for studies on skin ageing [61].

6.3 Disease Modelling

In the past, preclinical drug development for the treatment of skin diseases was mainly based on animal experiments. Current research is based on in vitro testing with human skin equivalents, either in addition to animal testing or increasingly alone. More recently, tissue culture technology has also been used to develop in vitro models of skin disease, particularly to promote research into cutaneous drugs while sparing experimental animals. The spectrum of disease models available ranges from inflammatory diseases to cancer [1, 8, 10, 11, 13]. There are two main methods for generating skin disease models, namely reconstruction with patient cells or the addition of genes and molecules to originally healthy bioengineered skin that act as disease triggers [64]

6.3.1 Atopic Dermatitis

Atopic dermatitis (AD) is a major public health problem, especially in children. It occurs due to genetic, environmental, and immunologic factors resulting in impaired skin epidermal barrier function. The aetiology of AD is complex and not yet fully understood [65, 66]. Both epidermal barrier impairment and systemic inflammation leading to an abnormal immune response result in a vicious cycle. Epidermal barrier defects allow for increased cutaneous permeability and consequently trigger inflammation and immune system activation. Conversely, skin inflammation leads to epidermal barrier defects with disruption of keratinocyte differentiation. During onset and progression, allergen-stimulated T helper type 2 (Th2) drive an immune response that triggers inflammatory signalling pathways via secreted pro-inflammatory cytokines. The positive feedback of signalling pathways leads to an overproduction of inflammatory cytokines in chronic AD. Th2-related cytokines IL-4 and IL-13 are involved in AD pathophysiology [65, 67]. As for the epidermis, both the organization and composition of proteins and lipids play an important role in the epidermal barrier function [66]. Although there are several published in vitro 3D models that mimic AD, none of them has considered all these pathophysiological features. In general, in vitro AD models are based on RHE models. Although it is not possible to reconstruct the genuine immunological response in them, they allow studying certain aspects of it. The induction of an RHE mimicking AD can be done by modifying the culture medium [10, 66].

Because of the many factors that can trigger AD, different groups have developed their own RHE-based AD models, each modifying the culture medium in different ways to trigger the disease. Nevertheless, no model reproduces all the features of AD. Since each model focuses on one or a few features of AD pathophysiology, it is important to select the correct model according to the goals of the analysis [10, 66]. The absence of fibroblasts, immune cells and nerve endings is one of the major downsides of the RHE model. However, inflammation of the skin may be mediated by immune cells, particularly Th2 cells, and may be mediated by neurogenic inflammation in AD [10, 66, 68]. Therefore, most models attempt to mimic inflammation by adding cytokines to the culture medium [68–72]. Keratinocytes are in constant contact with their environment and interact with the cells that make up their environment. These cells, particularly fibroblasts, play an important role in their differentiation and stratification. Explant skin models enable this communication and are obviously closer to "real" AD [73].

The establishment of AD-like properties in the RHE model was also observed after treatment with methyl-β-cyclodextrin, which causes cholesterol depletion in the plasma membrane, followed by IL-4, IL-13, and IL-15 exposure [74, 75]. Recently, a three-layer vascularized 3D model of AD revealed several hallmarks of the disease, including spongiosis and hyperplasia, early and terminal expression of differentiation proteins, as well as elevated levels of proinflammatory cytokines. In addition, the biofabricated AD tissue models demonstrated preclinical relevance for correcting the disease phenotype by testing different treatment strategies for AD from clinical trials [65].

6.3.2 Psoriasis

Psoriasis, a chronic autoimmune disease of the skin, causes scaly, erythematous lesions. Psoriasis may be associated with numerous comorbidities, including cardiovascular disease and arthritis [1, 10, 76]. The general classification for in vitro psoriasis models relies on whether they are 2D or 3D reconstructed skin substitutes. The monolayer (2D) models have been widely used. However, no interactions can be observed between the different cell subtypes that normally make up human skin, which has also led to increased interest in the 3D models [76]. The latter with replicated psoriasis-like conditions were developed to study the pathology of the disease and the prophylactic efficacy of drugs and identify new targets [1]. In vitro skin models were prepared from skin biopsies of psoriatic skin and healthy individuals or healthy skin only [77]. Psoriasis-like morphology was also achieved by cytokine stimulation [78]. In addition, cells from lesional and uninvolved skin of patients suffering from psoriasis vulgaris were used to develop full-thickness human disease models [79]. The pathophysiology of psoriasis involves a complex interaction between keratinocytes and immune cells [1, 13]. It has been shown that the combination of lesional keratinocytes and fibroblasts is quite representative of the morphology of human disease. Almost the same picture was obtained when lesional keratinocytes were combined with normal fibroblasts. Disease models that were considered particularly suitable were characterized by a thickening of the epidermis, with simultaneous overexpression of involucrin and underexpression of filaggrin and loricrin [77]. Mirroring the situation in patients, the chemokine receptor CXCR2 was overexpressed in the granular layer of keratinocytes in the full-thickness disease model. When the cells for the model were derived from lesional or non-lesional skin of psoriatic patients, pro-inflammatory genes were strongly expressed. This was especially true for tumor necrosis factor α (TNF-α), interferon-γ and IL-8. The results were completely different for tissues based on normal cells [79]. Using living skin equivalent, normal human keratinocytes behaved like typical psoriatic keratinocytes after stimulation with so-called psoriasis-associated cytokines (TNF-α and IL-1α) used alone or in combination. The associated increased expression of human β-defensins 2 (hBD-2), skin-derived anti-leucoproteinase (SKALP)/elafin, cyclokeratin 16 (CK16), IL-8 and TNF-α was prevented by retinoic acid. The antipsoriatic effect of cyclosporin could not be demonstrated as the model does not contain T-lymphocytes [78].

The cells that produce neuropeptides and hormones respond to neurotransmitters, helping to maintain homeostasis in the skin and throughout the body. In RHE models, these neuroimmune endocrine connections are absent [10]. Nevertheless, keratinocytes, melanocytes and cultured fibroblasts have been shown to produce corticotropin-releasing factor, pro-opiomelanocortin and their corresponding receptors [80]. In addition, the production of cortisol and corticosterone has also been demonstrated in keratinocytes, melanocytes, and cultured fibroblasts [81]. Therefore, in vitro models, both monolayers and bilayers can be useful for this type of study, as it is important to analyse the interactions at all levels [10]. These substitutes have allowed us to explore diseases such as psoriasis in greater depth, even if they are incomplete models lacking vascularization, innervation, and immune cells. On the other hand, surrogate models are made with only keratinocytes and fibroblasts, as the gene expression profile analysis is reliable since no other cell types are present in the model. These models are a major advance in developing new treatments for psoriasis [10, 76].

6.3.3   Cutaneous Candidosis

Cutaneous candidosis (CC), also known as cutaneous candidiasis, is a fungal infection caused by the dimorphic yeast *Candida albicans* (*C. albicans*). The use of RHE offers the basis for an in vitro model of human cutaneous candidiasis. Morphologically, RHE partially resembles normal human epidermis, but its barrier function does not adequately reflect the situation in vivo. Modifying the culture medium (as described in Sect. 6.2.1) could solve this problem. In addition, the protective *stratum corneum* necessitates the application of an irritant chemical (sodium dodecyl sulfate, SDS, or sodium lauryl sulfate, SLS) before inoculation to ensure that the relevant damage would occur. Alternatively, a mechanical insult can be performed with a hypodermic needle [13, 82]. An in vitro model of CC was also developed by infecting SLS-treated RHE with *C. albicans* blastospores, which caused hyperkeratosis, scaling, and superficial keratin breakdown. Secreted aspartyl proteinases were responsible for the infection [82].

Using 3D skin models supplemented with immune cells (CD4$^+$ T cells), it was possible to define a new role for fibroblasts in the dermis and identify a minimal set of cell types for protecting the skin from *C. albicans* invasion. This model demonstrated that communication between T cells and fibroblasts plays an important role in protecting against CC. As shown, dermal fibroblasts were able to integrate signals from the pathogen and CD4$^+$ T cells, which were critical for initiating an antimicrobial response by the fibroblasts. These results underline a central function of dermal fibroblasts for skin protection and open new possibilities for treating infectious diseases [83].

The RHE-based cutaneous candidiasis model allowed the characterisation of topical econazole formulations. In this study, a liposomal formulation proved superior to the conventional cream formulation [84]. In particular, hyperkeratosis was reduced, as evidenced by the only moderate detachment of the *stratum corneum* and less pronounced vacuolization. The higher in vitro activity of the liposomal formulation was reflected in a higher clinical efficacy [85].

6.3.4   Bacterial Infection

Engineered skin equivalents have rarely been used to create full-scale bacterial skin and disease models. Nevertheless, 3D human models have proven to be a valuable tool for one step closer to uncovering the causative factors that influence skin barrier and microbial colonization. In particular, these in vitro skin models have been used to characterize the interaction of skin tissue with *S. aureus*, which can increase its density by several orders of magnitude within a few days [13, 86, 87]. Using RHE, it was shown, for example, that knockdown of FLG, which is associated with a compromised lipid envelope and thus impaired barrier function, is associated with increased colonization by *S. aureus* [88].

To limit or even prevent this biocontamination, it is essential to understand the mechanisms involved in adhesion between the microorganism and the skin. Inert substrates (polymers, metals, glass) can change these properties that have the potential for biocontamination. Surface roughness or topography may also play a role in microbial adhesion. In this regard, the commercial RHE model EpiSkin™ infected with *S. aureus* (a potentially pathogenic bacteria of the skin) and *S. epidermidis* (a saprophytic bacteria of the skin) served as a model biological substrate. It was used to highlight the respective influence of physicochemical interactions and roughness involved in the first part of biocontamination of the biological substrate and to evaluate the influence of roughness on the initiation of this bioadhesive phenomenon. With respect to the degree of biocontamination of the different substrates employed (Episkin™ or medical grade stainless steel), a higher degree of adhesion was observed for *S. aureus* than for *S. epidermidis*, indicating the important participation of van der Waals and Lewis acid–base interactions in the biocontamination of the Episkin™ biological substrate. To limit microbial adhesion and reduce adhesive binding between the microorganisms and the skin surface (in the absence of any electrostatic interactions) or the stainless steel, it would be preferable to make this substrate hydrophobically apolar by a suitable surface treatment [89].

Epidermis models also help evaluate the therapeutic role of dressings in methicillin-resistant *S. aureus* (MRSA) and *C. albicans* colonization. Hydrocolloid dressings were compared for their antimicrobial activity, whether or not they contained silver particles. Treatment with both dressings reduced the number of *C. albicans* or MRSA cells, with the silver-based dressing demonstrating more effective antimicrobial activity. The superiority of the silver-based dressing over the silver-free dressing was revealed in this study by increased antimicrobial activity and decreased invasion by bacterial and fungal pathogens. This improved antimicrobial efficacy against the pathogens was associated with less toxic effects on the epithelium, as shown by histological alterations [90].

6.3.5 Skin Cancer

Skin cancer ranges from basal cell carcinoma, squamous cell carcinoma to melanoma and is one of the most common cancers in Caucasian people worldwide. Skin cancer is often associated with photodamaged tissues, and constructed 3D skin cancer models have proven useful in elucidating the dermo-epidermal and tumour-stroma interactions in skin cancer [1, 8]. The most common in vitro skin cancer model is the melanoma model. When melanocytes grow in the 3D reconstructed skin, they exhibit physiological features of melanocyte homeostasis and melanoma progression observed in the skin of human patients. When melanoma cells are incorporated into the 3D skin reconstruct, they exhibit characteristics consistent with the aggressiveness of the human melanoma patient. Critical to the success of this model is the ability to grow viable melanocytes, melanoma cells, keratinocytes, and fibroblasts for use in 3D reconstructed skin. Melanocytes can be derived from human skin but can also come from ESCs or iPSCs [91]. Hill and his group [92] recapitulated an invasive melanoma using rat melanoma cells in a 3D skin equivalent, demonstrating that it is currently possible to incorporate primary cells into a 3D skin model in vitro. A fully humanized melanoma 3D skin model was constructed using an inert porous scaffold (Alvetex®, Reinnervate Ltd., Reprocell Group), into which human fibroblasts were incorporated to generate the dermis. This system has eliminated the need to use collagen from bovine or rat tails, which are not representative of the normal skin microenvironment. The scaffold allows the 3D growth of fibroblasts, which are stimulated to produce their own ECM components and form a stable dermis similar to humans. Melanoma cells implanted in the epidermis could penetrate through the basement membrane into the dermis, mirroring early tumor invasion in vivo. Comparison of a 3D melanoma skin model with melanoma in situ and metastatic melanoma demonstrated that this 3D melanoma model accurately simulated the features of the disease pathology, providing a physiologically representative model of the early radial and vertical growth phase of melanoma [92]. Alternatively, the multicellular tumour spheroid (MCTS) can be used as a 3D melanoma model. MCTS not only provide relevant information on the intricate cell–cell and cell–matrix interactions, hypoxia, and tumour metabolism but are also reliable in studying tumour progression and invasion. Furthermore, these can be used in evaluating the effects of drugs on cell migration [93]. In an organotypic skin-melanoma spheroid model (Fig. 6.2), melanoma MCTS are integrated into the dermal compartment rather than seeded with keratinocytes over collagen-derived dermal equivalents. Consequently, MCTS in the dermis closely resemble in vivo cutaneous melanoma metastases, resulting in a large dermal melanoma nest [94].

Melanoma grows in a 3D spatial conformation in which cells are subjected to a heterogeneous exposure to oxygen and nutrients. Moreover, cell–cell and cell–matrix interactions play a critical role in tumour pathobiology and response to therapeutic agents. The unique architecture and composition of the 3D melanoma model allow a thorough investigation of the autocrine and paracrine loops between melanoma cells, keratinocytes and fibroblasts [91, 93, 95].

Hanging drop cell culture is an alternative method for generating MTCS. This technique is useful to propagate cells in a 3D system, but it is necessary to transfer

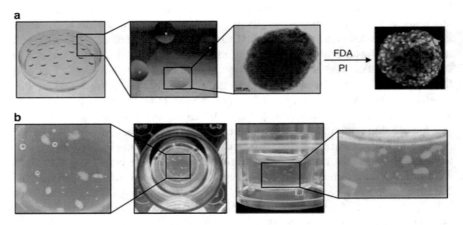

Fig. 6.2 Generation of the organotypic melanoma-spheroid skin model. **a** Melanoma cells were placed on the lid of a non-adhesive petri dish and incubated for 15 days deriving melanoma spheroids. Live/death staining of melanoma spheroids. **b** Melanoma spheroids were collected, carefully mixed with primary fibroblasts and added to collagen I to from the dermal compartment of the full thickness skin equivalent. Reproduced from [94] with permission from Springer Nature Limited

MTCS to another plate for further studies. Bourland and co-workers [96] have introduced melanoma spheroids cultured by the hanging drop method into a human skin equivalent model that exhibits microvascularization of blood and lymphatic vessels. Capillary networks are involved in tumour progression and are particularly important for modelling melanoma. The addition of endothelial cells, such as HuMVEC or HUVEC, contributed to a more complex tumour microenvironment. Also, lymphatic endothelial cells (LECs) can be isolated from HuMVECs and incorporated into 3D constructs that can assemble into lymphatic capillaries distinct from those formed by blood endothelial cells (BEC) in a collagen gel. HuMVECs were incorporated into cells of the dermal compartment to recapitulate the formation of blood and lymphatic capillaries, followed by the addition of melanoma spheroids to the formed tissue. The latter were rapidly integrated into the epidermal, dermo-epidermal and dermal regions of the reconstructed skin equivalents. The most important features of this melanoma skin model were the reproduction of the original tissue characteristics (structure and cell types) and the response of the spheroids to known anti-melanoma therapeutics [96].

In addition, 3D immunoskin models have also been developed that contain melanoma cells, autologous T cells, and fibroblasts that may be allogeneic. This approach allows visualization and quantification of T cell-mediated killing of tumour cells [91, 97]. To create such a four-layer 3D immunoskin model, human fibroblasts are suspended in collagen, which is the first layer. Melanoma cells are then layered on top of the collagen/fibroblast matrix and grow overnight (the second layer). They are then covered with a cell-free collagen layer (the third layer). Next, an equal number of immune T cells (antitumor reactive cytotoxic T cells) are mixed with fibroblasts in collagen and layered on top, forming the fourth layer [91, 98]. A major advantage of 3D immune assays is the "natural" setting, in which malignant, stromal, and immune cells migrate to each other through collagen layers. Since every coin has two sides, its disadvantage is that autologous pairs of T cells and tumour cells are needed, which are easy to make for melanomas but more challenging for epithelial tumours [91].

6.4 Product Testing

In addition to various vital functions, human skin also provides a potential site for the transport of functional, active drugs/reagents/ingredients into the skin (topical administration) and/or the body (transdermal administration) [99, 100]. Any novel substance used in the composition of consumer products must first undergo a comprehensive toxicological evaluation. The use of animal testing continues to be strongly questioned and rightly criticized on both ethical and scientific grounds [101]. Compared to other mammals, human skin has unique features (pigmentation, dermis development, adipose tissue, and distribution of skin appendages) that make comparisons of skin behaviour between species inaccurate [2, 102]. Thus, the penetration and absorption of chemicals and substances into human skin cannot be fully predicted by studying these phenomena on the skin of other mammals because the structure and organization are distinct. This is particularly important in developing skin care products, as many companies currently use animal model testing in this process [2]. Current skin model development efforts include adding dermal layer components to create a 3D model that can be used to study permeability and absorption. Efforts are also being made to improve the limitations of current models, such as further evolving their barrier functions. Indeed, it was found that there is a difference in lipid organization between reconstructed skin equivalents and native human skin tissue, which presumably contributes to the 5–50-fold higher penetration rate observed in human skin models for most of the tested substances [103].

The cosmetics and skin care industry are among the fastest growing markets globally due to the continuous development of beauty and hygiene products [2]. Makeup products, also called colour cosmetics, are usually applied for a short time and removed or cleansed after use. Skin enhancing products such as moisturizing products, formulations enriched with bioactive molecules (OCTs), tanning creams or bleaching agents are intended to have a more lasting effect on the appearance of the skin and may remain in the skin for a longer period. For this reason, companies developing such products need skin mimicking products to determine what concentration of certain compounds or chemicals will achieve the product's goal without harming the consumer and meeting regulatory requirements. Skin care products are another important application for skin models used in product development and evaluation. Unlike makeup products and cosmetics, skin care products are meant for cleansing, moisturizing and refreshing. These products come in the form of lotions, creams, scrubs, and serums. Common skin cleansers are made of synthetic detergents and have a low alkaline pH, while most soaps are anionic surfactants with higher pH values. Due to the nature of these products, a balance between hygiene and protection of the *stratum corneum* barrier must be struck. Therefore, testing such products is crucial to determine the exact effect on the lipid composition of the *stratum corneum* [104].

Pharmaceutical products present greater challenges. Traditional 2D skin models (monolayer cultures of keratinocytes or co-cultures of keratinocytes with immune cells and dermal fibroblasts) are well established and easy to use. However, they cannot replicate the complex 3D cell–cell and cell–matrix interactions present in the body, limiting their accuracy in predicting the complicated effects of drug metabolism on actual skin [105]. To tackle these limitations, in vitro 3D skin models with cells cultured in ECM-like materials are used in drug testing, which helps to study the physical distribution of the drug. Nevertheless, most conventional 3D skin models are still unable to reconstruct the multicellular complexity of human skin tissue fully. Among the reasons for this are their weak barrier properties and lack of blood vessels and skin appendages (sweat glands and hair follicles), limiting their applicability to address the physical distribution of the drug. The development of novel in vitro skin models, "skin-on-a-chip" (SoC), can bridge the gap between currently used 3D skin models and the in vivo situation and allow the assessment of the penetration, metabolism and effect of a topically and/or transdermally applied drugs [2, 105].

Skin testing procedures for new chemicals and finished products are evolving rapidly in the face of technological advances and political pressures. Engineered skin substitutes are currently used in the cosmetic industry and pharmacological research as increasingly more reliable model systems to identify the irritant, toxic, or corrosive properties of chemical substances that come into contact with human skin [106].

6.4.1 Percutaneous Penetration

Evaluation of percutaneous permeation is key to the successful development of new formulations for human use. It is also an important quality control measure to ensure batch-to-batch consistency in the pharmaceutical industry [100]. In particular, permeation and safety evaluation are key points for developing new cosmetic or drug formulations for human use. Although the use of human skin is the gold standard, human skin, like animal models, is subject to high variability, making the results of such a study less accurate [2, 100, 107]. Therefore, in vitro skin models have become a useful alternative to both human and animal skin models, as they are more reproducible and allow for more consistent results.

Diffusion through the skin, controlled by the outermost layer, the *stratum corneum*, can be thought diffusion through a passive membrane. Researchers often use an in vitro protocol with a membrane sandwiched between two compartments in skin permeation studies. One compartment contains a drug formulation (the donor), and the other compartment holds a receptor solution that provides sink conditions (essentially zero concentration) [108]. The parallel artificial membrane permeation assay (PAMPA) using poly(2-acrylamide-2-methyl-1-propanamide) is a rapid screening of passive transport across a given membrane and simulates epidermal barrier function [109, 110]. A skin-PAMPA artificial membrane introduced by Ottaviani and colleagues [110] showed comparable permeability to human skin for various substances, confirming the efficacy of using a synthetic skin membrane for permeability studies [110]. Sinkó and his group went one step further and created the skin-PAMPA membrane containing synthetic ceramides to replace the naturally present ceramides in the *stratum corneum* [111]. With Tsinman [112], Sinkó refined the skin-PAMPA to predict skin penetration and screen topical formulations (silicone-based gel, silicone, and acrylic copolymer). Another model designed to mimic biological barrier cells is a phospholipid vesicle-based barrier assay (PVPA) consisting of a dense layer of liposomes. In its original form, it served as a screening model for intestinal permeability [113]. Further improvement of PVPA led to novel skin PVPA models that closely resemble the epidermal skin barrier. Two modifications of the skin PVPA model were developed to estimate skin penetration:

(i) using liposomes of cholesterol and egg phosphatidylcholine,
(ii) using the major lipid classes found in the skin (ceramide, cholesterol, FFA and cholesteryl sulfate, and egg phosphatidylcholine).

Furthermore, the barrier function of the skin-PVPA model could be modified in a controlled manner, and it was shown that the PVPA barriers could mimic various biological barriers by exchanging the lipid components in the barriers [114, 115]. Subsequently, the PVPA model mimicking the *stratum corneum* was adapted to evaluate the effect of the vesicle carrier on skin penetration [116].

Some reconstructed skin models are produced in-house in various companies. In contrast, others are commercially available (EpiSkin™, SkinEthic™, Apligraf® and EpiDerm™). The literature reports both options as suitable candidates for in vivo and ex vivo skin models to evaluate absorption [100, 101]. Several studies have compared in vitro 3D skin models with animal (rat and pig) and human skin. Their general conclusions were that the RHE models, in particular SkinEthic™, were significantly more permeable than the ex vivo skins. However, the ranking of permeation of compounds through pig skin and RHEs was similar to that through the human epidermis. The higher permeation, and thus overestimation of skin absorption, using RHEs is consistent with the incomplete barrier found in these models. The deficiencies in the barrier are caused by lower concentrations of FFAs and hydrophilic ceramide fractions, as well as the expression of cytokines and GFs that lead to hyperproliferation of epidermal cells [117]. Interestingly, they did not observe the expected improvement in reproducibility with the RHEs compared to ex vivo skin [118–120].

Routine protocols for penetration studies have been developed to screen new chemicals and mixtures [101]. Using the EpiSkin™ model, it has been demonstrated that percutaneous assays can be easily performed without the need for a specialised or complex device such as the Franz cell [121]. This system facilitates the development of screening assays to evaluate skin penetration of compounds with high reliability and throughput. The permeation of essential oils was evaluated using SkinEthic™ RHE to determine the kinetics of the release of active ingredients from cosmetic formulations. Diffusion and quantification of selected terpenes proved that the method is sensitive, simple and reproducible, indicating its convenience for evaluating the safety and quality of formulations [122]. Similarly, EpiSkin™ tissues were used to evaluate the absorption of vitamins C and E from topical microemulsions and select the best formulation capable of enhancing the penetration of the tested vitamins into the tissue [123].

Percutaneous penetration is greatly influenced by the physiological and lipophilic characteristics of substances (the lower the molecular weight and the lower the water solubility of the substance, the more easily it can penetrate the dermis). Penetration is enhanced when the active ingredients are diluted in water (25% for lipophilic active ingredients and 10–20% for hydrophilic ones) [124]. In addition, tissue hydration has been shown to increase the transdermal delivery of substances [108].

Determination of dermal absorption in the presence of a damaged skin barrier is particularly important for products such as sunscreens, creams for sun treatments, creams for irritated or diseased skin, which should only be applied to sore or irritated skin. Damaged skin is likely to have a lower barrier function than healthy skin. The most common methods of compromising the epithelium are tape stripping and cyanoacrylate stripping. In tape stripping, a strip with the attachment is placed on the skin and pressed on with a rubber roller. The roller is moved back and forth 10 times, and then the strip is removed from the skin in one swift motion. Extraction of 10 consecutive strips can alter the properties of the barrier in the in vitro model and cause corresponding damage to the skin barrier in vivo. In cyanoacrylate stripping, an adhesive is applied to the skin, and an adhesive tape is placed on the adhesive under pressure (10 N) for 5 min, then the tape is removed [10, 125]. When assessing risk using in vitro skin models to estimate systemic exposure to substances, it is important to consider whether the product is applied to intact or damaged skin, as systemic exposure may differ by up to tenfold depending on barrier properties and substance composition [10].

6.4.2 Skin Corrosion/irritation

The manufacture, transportation, and marketing of chemicals and finished products require prior toxicological evaluation and assessment of skin reactivity (corrosion and irritation) that may occur with intentional or accidental skin exposure [106]. Skin corrosion or dermal corrosion tests evaluate the potential of a substance to cause irreversible damage to the skin (visible necrosis through the epidermis and into the dermis) after a test substance has been applied for a predefined time (between 3 min and 4 h). Skin irritation or dermal irritation is defined as reversible skin damage after applying a test substance for up to 4 h. Evaluation of the skin irritation potential of chemicals or products used in the pharmaceutical or cosmetic industry is required [101]. Depending on the severity of the skin reactions (erythema, oedema, necrotic changes), the speed of their occurrence, and their persistence and reversibility, classifications of the risk of skin corrosion and irritation are made [106]. The Draize test has been used for skin irritation and corrosion testing for several decades [126]. Because it is very uncomfortable and painful for the animal (rabbit), it was one of the animal tests that was banned with the 2013 deadline [101]. In the test, a topical product is applied once to the intact skin of an animal, such as a rabbit. The test substance is removed after approximately 4 h and immediately assessed. Exposure to the test substance is continued at specified intervals over several days. This method is not only questionable for animal welfare reasons but also leads to very different results [2, 126].

Skin corrosivity differs from skin irritation in two important respects. Corrosive skin reactions generally occur rapidly after chemical exposure and are irreversible. In addition, the main direct processes leading to chemical corrosivity are thought to be physicochemical rather than the result of inflammatory biological processes. However, inflammation is certainly a consequence of corrosive skin events [106]. Initial attempts to develop a screening test for skin corrosivity built on this hypothesis and examined the effects of chemical exposure on skin barrier function by assessing changes in the resistance of exposed skin to the transmission of electric current [106, 127]. In the early 1990s, under the auspices of the European Center for the Validation of Alternative Methods (ECVAM), a program was initiated to develop and validate alternative methods for the assessment of skin corrosion [106]. During the ECVAM-sponsored skin corrosivity validation study program (from early 1995 to October 1997), only the Episkin™ assay system met all criteria. It demonstrated acceptable intra- and inter-laboratory reproducibility and was able to discriminate corrosive from non-corrosive chemicals with acceptable under- or overestimation rates. In addition, the SkinEthic™ RHE model was also accepted as a means of discriminating between corrosive and non-corrosive reference chemicals with an accuracy of 93% [106, 128].

The development of in vitro tests for skin irritation is complicated because skin irritation covers a wide range of severity, from near corrosivity on the one hand to mild cumulative or sensory-only irritation on the other [106]. Several other laboratories worldwide have been active in developing skin-equivalent culture systems to evaluate the skin irritation potential of both ingredients and finished products. The Epiderm™ culture system has been used in studies of surfactant and surfactant-containing formulations concerning clinical irritation profiles [129, 130] and to compare chemical skin irritants and skin allergens [131]. It has also been used as a test system for in vitro evaluation of antagonism in mixed surfactant systems. Various research groups have used either Episkin™ or non-commercial skin equivalent cultures to investigate the irritant properties of various cosmetic formulations [132, 133]. These groups have achieved good correlations between their in vitro test results and comparable in vivo skin irritation data sets. A validated alternative to the Draize test for skin irritation was the SkinEthic™ test method. The test consisted of a 42-min topical application of test substances followed by a 42-h post-incubation period. This method showed good reproducibility in three independent laboratories and provided reliable results with an accuracy of 85% [134]. The overall accuracy of the SkinEthic™ RHE test method led to its recognition by the ECVAM Scientific Advisory Committee as the only verified replacement method for the Draize rabbit in vivo test in the evaluation of irritancy of test substances [2, 134]. In addition, the SkinEthic™ RHE model was also accepted as a means of discriminating between corrosive and non-corrosive reference chemicals with an accuracy of 93% [128].

6.4.3   Genotoxicity

Genotoxicity is another problem in testing the efficacy of cosmetic and pharmaceutical products. In short, genotoxicity is the destructive effect on the genetic material. Any substance that has a genotoxic effect is called a genotoxin. Genotoxins include cancer, mutation and congenital disability causing substances [2]. Genotoxicity and mutagenicity assessment is an important and early step in the safety evaluation of chemicals for industrial development and regulatory purposes. Although many in vitro tests are routinely used and accepted by regulatory agencies, their accuracy in predicting mutagenic/genotoxic potential in humans is often questioned [101]. Improving current in vitro tests for genotoxicity is an ongoing task for genetic toxicologists. Another challenge is how to deal with positive in vitro results that have been shown not to predict genotoxicity or carcinogenicity potential in rodents or humans. Positive in vitro results that are not predictive of genotoxic/carcinogenic potential in rodents or humans may trigger unnecessary in vivo follow-up testing and require significant time and human resources from both regulatory agencies and the industry. Due to the resources involved in clarifying positive results of standard in vitro tests, companies often refrain from using such ingredients, thereby losing potentially safe and useful candidate chemicals [135]. In vivo genotoxicity testing was declared impractical on a large scale. This led to a ban on the test method for cosmetic ingredients in Europe and other chemical evaluation programs since 2009 [2, 136]. To replace in vivo testing methods, researchers have sought to develop in vitro genotoxicity tests using reconstructed human skin models such as EpiSkin™ and EpiDerm™ [2, 101, 136]. Specific protocols for performing the comet assay (or the single cell gel electrophoresis assay, which is a sensitive technique for detecting DNA damage at the level of the individual eukaryotic cell [137]) on EpiSkin™ were developed using both UV exposure and/or reference chemicals (lomefloxacin, 4-nitroquinoline N-oxide) as stressors [101, 138]. Pretreatment with the sunscreen Mexoryl SX® was able to significantly reduce the extent of UV-induced DNA damage measured from the comet signal. A second approach (Fig. 6.3) was developed using the barrier function and metabolic capacities of the EpiSkin™ model to drive the response to a topically applied test chemical and its potential metabolites. In this approach, a specific co-culture system was performed with a target lymphoma cell line (L5178Y) under the epidermis to evaluate micronucleus induction after topical application of the test chemical. This approach aims to improve the relevance of exposure conditions for testing further products applied to the skin [138].

The obtained results illustrate the feasibility and relevance of the in vitro reconstructed human skin model as a biologically active barrier for in vitro clastogenicity

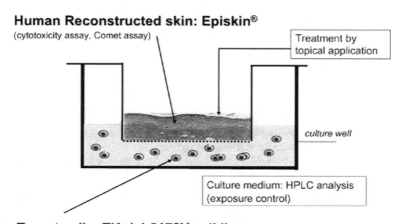

Fig. 6.3 Genotoxicity testing and human reconstructed skin epidermis: the Episkin®-L5178Y cell co-culture system. This system was used to study DNA damage using the comet assay on cells in the reconstructed epidermis and to evaluate micronuclei in target cells cultured under Episkin. Reproduced from [138] with permission from Elsevier B.V

assays on cultured cells. However, several limitations, such as the metabolism of the EpiSkin™ assay system, still need to be addressed before this system can be used for routine safety assessment of chemical agents applied to the skin. Another approach currently being validated is to perform the micronucleus test, and the comet assay directly on 3D reconstructed human skin models [135].

These various projects on reconstructed epidermis and skin have helped improve the predictive power of in vitro assays. Optimization of protocols has resulted in improved specificity of the micronucleus test, such that more than 60% of irrelevant positive findings have been avoided by using the optimized methods. Both the micronucleus test and the comet assay in reconstructed skin models are not considered as stand-alone assays but are ideally suited for non-animal genotoxicity testing to confirm positive in vitro genotoxicity test results [101].

References

1. Choudhury S, Das A (2020) Advances in generation of three-dimensional skin equivalents: pre-clinical studies to clinical therapies. Cytotherapy
2. Suhail S, Sardashti N, Jaiswal D, Rudraiah S, Misra M, Kumbar SG (2019) Engineered skin tissue equivalents for product evaluation and therapeutic applications. Biotechnol J 14(7):1900022
3. Slivka SR, Landeen LK, Zeigler F, Zimber MP, Bartel RL (1993) Characterization, barrier function, and drug metabolism of an in vitro skin model. J Investig Dermatol 100(1):40–46
4. Nachman M, Franklin S (2016) Artificial skin model simulating dry and moist in vivo human skin friction and deformation behaviour. Tribol Int 97:431–439
5. Bhushan B, Tang W (2011) Surface, tribological, and mechanical characterization of synthetic skins for tribological applications in cosmetic science. J Appl Polym Sci 120(5):2881–2890
6. Franklin S, Baranowska J, Hendriks C, Piwowarczyk J, Nachman M (2017) Comparison of the friction behavior of occluded human skin and synthetic skin in dry and moist conditions. Tribol Trans 60(5):861–872
7. Mohd Noor SNA, Mahmud J (2014) A review on synthetic skin: materials investigation, experimentation and simulation. Adv Mater Res, Trans Tech Publ, pp 858–866
8. Randall MJ, Jüngel A, Rimann M, Wuertz-Kozak K (2018) Advances in the biofabrication of 3D Skin in vitro: healthy and pathological models. Front Bioeng Biotechnol 6:154
9. Savoji H, Godau B, Hassani MS, Akbari M (2018) Skin tissue substitutes and biomaterial risk assessment and testing. Front Bioeng Biotechnol 6:86
10. Sanabria-de la Torre R, Fernández-González AFV, Quiñones-Vico MI, Montero-Vilchez T, Arias-Santiago S (2020) Bioengineered skin intended as in vitro model for pharmacosmetics, skin disease study and environmental skin impact analysis. Biomedicines 8(11):464
11. Sarkiri M, Fox SC, Fratila-Apachitei LE, Zadpoor AA (2019) Bioengineered skin intended for skin disease modeling. Int J Mol Sci 20(6):1407
12. Yu JR, Navarro J, Coburn JC, Mahadik B, Molnar J, Holmes JH IV, Nam AJ, Fisher JP (2019) Current and future perspectives on skin tissue engineering: key features of biomedical research, translational assessment, and clinical application. Adv Healthcare Mater 8(5):1801471
13. Semlin L, Schäfer-Korting M, Borelli C, Korting HC (2011) In vitro models for human skin disease. Drug Discovery Today 16(3–4):132–139
14. Turner NJ, Badylak SF (2015) The use of biologic scaffolds in the treatment of chronic nonhealing wounds. Adv Wound Care 4(8):490–500
15. Spear M (2011) Skin grafts: indications, applications and current research

16. Shevchenko RV, James SL, James SE (2010) A review of tissue-engineered skin bioconstructs available for skin reconstruction. J R Soc Interface 7(43):229–258
17. Converse JM, Smahel J, Ballantyne DL Jr, Harper AD (1975) Inosculation of vessels of skin graft and host bed: a fortuitous encounter. Br J Plast Surg 28(4):274–282
18. Markeson D, Pleat JM, Sharpe JR, Harris AL, Seifalian AM, Watt SM (2015) Scarring, stem cells, scaffolds and skin repair. J Tissue Eng Regen Med 9(6):649–668
19. Metcalfe AD, Ferguson MW (2007) Tissue engineering of replacement skin: the crossroads of biomaterials, wound healing, embryonic development, stem cells and regeneration. J R Soc Interface 4(14):413–437
20. Peck M, Molnar J, Swart D (2009) A global plan for burn prevention and care. Bull World Health Organ 87:802–803
21. Gurtner GC, Werner S, Barrandon Y, Longaker MT (2008) Wound repair and regeneration. Nature 453(7193):314–321
22. Hu MS, Maan ZN, Wu J-C, Rennert RC, Hong WX, Lai TS, Cheung AT, Walmsley GG, Chung MT, McArdle A (2014) Tissue engineering and regenerative repair in wound healing. Ann Biomed Eng 42(7):1494–1507
23. Thornton JF, Gosman A (2004) Skin grafts and skin substitutes. Sel Read Plast Surg 10(1):1–24
24. Choi J, Lee EH, Park SW, Chang H (2015) Regulation of transforming growth factor β1, platelet-derived growth factor, and basic fibroblast growth factor by silicone gel sheeting in early-stage scarring. Arch Plast Surg 42(1):20
25. Oualla-Bachiri W, Fernández-González A, Quiñones-Vico MI, Arias-Santiago S (2020) From grafts to human bioengineered vascularized skin substitutes. Int J Mol Sci 21(21):8197
26. Jones I, Currie L, Martin R (2002) A guide to biological skin substitutes. Br J Plast Surg 55(3):185–193
27. Freyman T, Yannas I, Gibson L (2001) Cellular materials as porous scaffolds for tissue engineering. Prog Mater Sci 46(3–4):273–282
28. Uchino T, Takezawa T, Ikarashi Y (2009) Reconstruction of three-dimensional human skin model composed of dendritic cells, keratinocytes and fibroblasts utilizing a handy scaffold of collagen vitrigel membrane. Toxicol In Vitro 23(2):333–337
29. Wang H-M, Chou Y-T, Wen Z-H, Wang Z-R, Chen C-H, Ho M-L (2013) Novel biodegradable porous scaffold applied to skin regeneration. PLoS ONE 8(6):e56330
30. Llames SG, Del Rio M, Larcher F, García E, García M, Escamez MJ, Jorcano JL, Holguín P, Meana A (2004) Human plasma as a dermal scaffold for the generation of a completely autologous bioengineered skin. Transplantation 77(3):350–355
31. Sakamoto M, Morimoto N, Ogino S, Jinno C, Taira T, Suzuki S (2016) Efficacy of gelatin gel sheets in sustaining the release of basic fibroblast growth factor for murine skin defects. J Surg Res 201(2):378–387
32. Xu S, Sang L, Zhang Y, Wang X, Li X (2013) Biological evaluation of human hair keratin scaffolds for skin wound repair and regeneration. Mater Sci Eng: C 33(2):648–655
33. Sun G, Zhang X, Shen Y-I, Sebastian R, Dickinson LE, Fox-Talbot K, Reinblatt M, Steenbergen C, Harmon JW, Gerecht S (2011) Dextran hydrogel scaffolds enhance angiogenic responses and promote complete skin regeneration during burn wound healing. Proc Natl Acad Sci 108(52):20976–20981
34. Loan F, Cassidy S, Marsh C, Simcock J (2016) Keratin-based products for effective wound care management in superficial and partial thickness burns. Burns 42(3):541–547
35. Webber MJ, Khan OF, Sydlik SA, Tang BC, Langer R (2015) A perspective on the clinical translation of scaffolds for tissue engineering. Ann Biomed Eng 43(3):641–656
36. Haldar S, Sharma A, Gupta S, Chauhan S, Roy P, Lahiri D (2019) Bioengineered smart trilayer skin tissue substitute for efficient deep wound healing. Mater Sci Eng: C 105:110140
37. Schmidt FF, Nowakowski S, Kluger PJ (2020) Improvement of a three-layered in vitro skin model for topical application of irritating substances. Front Bioeng Biotechnol 8:388
38. Lee MH, Arcidiacono JA, Bilek AM, Wille JJ, Hamill CA, Wonnacott KM, Wells MA, Oh SS (2010) Considerations for tissue-engineered and regenerative medicine product development prior to clinical trials in the United States. Tissue Eng Part B Rev 16(1):41–54

39. Mestas J, Hughes CC (2004) Of mice and not men: differences between mouse and human immunology. J Immunol 172(5):2731–2738
40. Seok J, Warren HS, Cuenca AG, Mindrinos MN, Baker HV, Xu W, Richards DR, McDonald-Smith GP, Gao H, Hennessy L (2013) Genomic responses in mouse models poorly mimic human inflammatory diseases. Proc Natl Acad Sci 110(9):3507–3512
41. Kim AM, Tingen CM, Woodruff TK (2010) Sex bias in trials and treatment must end. Nature 465(7299):688–689
42. Shultz LD, Ishikawa F, Greiner DL (2007) Humanized mice in translational biomedical research. Nat Rev Immunol 7(2):118–130
43. Van der Veen VC, van der Wal MB, van Leeuwen MC, Ulrich MM, Middelkoop E (2010) Biological background of dermal substitutes. Burns 36(3):305–321
44. Catalano E, Cochis A, Varoni E, Rimondini L, Azzimonti B (2013) Tissue-engineered skin substitutes: an overview. J Artif Organs 16(4):397–403
45. Bailey AM, Arcidiacono J, Benton KA, Taraporewala Z, Winitsky S (2015) United States Food and Drug Administration regulation of gene and cell therapies, Regulatory Aspects of Gene Therapy and Cell Therapy Products, pp 1–29
46. Mendicino M, Fan Y, Griffin D, Gunter KC, Nichols K (2019) Current state of US Food and Drug Administration regulation for cellular and gene therapy products: potential cures on the horizon. Cytotherapy 21(7):699–724
47. Mieremet A, García AV, Boiten W, van Dijk R, Gooris G, Bouwstra JA, El Ghalbzouri A (2019) Human skin equivalents cultured under hypoxia display enhanced epidermal morphogenesis and lipid barrier formation. Sci Rep 9(1):1–12
48. Mieremet A, Helder R, Nadaban A, Gooris G, Boiten W, El Ghalbzouri A, Bouwstra JA (2019) Contribution of palmitic acid to epidermal morphogenesis and lipid barrier formation in human skin equivalents. Int J Mol Sci 20(23):6069
49. Yuki T, Komiya A, Kusaka A, Kuze T, Sugiyama Y, Inoue S (2013) Impaired tight junctions obstruct stratum corneum formation by altering polar lipid and profilaggrin processing. J Dermatol Sci 69(2):148–158
50. Yuki T, Tobiishi M, Kusaka-Kikushima A, Ota Y, Tokura Y (2016) Impaired tight junctions in atopic dermatitis skin and in a skin-equivalent model treated with interleukin-17. PLoS ONE 11(9):e0161759
51. Huth S, Schmitt L, Marquardt Y, Heise R, Lüscher B, Amann PM, Baron JM (2018) Effects of a ceramide-containing water-in-oil ointment on skin barrier function and allergen penetration in an IL-31 treated 3D model of the disrupted skin barrier. Exp Dermatol 27(9):1009–1014
52. Geer DJ, Swartz DD, Andreadis ST (2002) Fibrin promotes migration in a three-dimensional in vitro model of wound regeneration. Tissue Eng 8(5):787–798
53. Garcia M, Escamez MJ, Carretero M, Mirones I, Martinez-Santamaria L, Navarro M, Jorcano JL, Meana A, Del Rio M, Larcher F (2007) Modeling normal and pathological processes through skin tissue engineering, Molecular Carcinogenesis: Published in cooperation with the University of Texas MD Anderson Cancer Center 46(8):741–745
54. Healy C, Boorman J (1989) Comparison of EZ Derm and Jelonet dressings for partial skin thickness burns. Burns 15(1):52–54
55. Küchler S, Wolf NB, Heilmann S, Weindl G, Helfmann J, Yahya MM, Stein C, Schäfer-Korting M (2010) 3D-wound healing model: influence of morphine and solid lipid nanoparticles. J Biotechnol 148(1):24–30
56. Korting HC, Schäfer-Korting M (2010) Carriers in the topical treatment of skin disease. Drug Deliv., 435–468
57. Schäfer-Korting M, Mehnert W, Korting H-C (2007) Lipid nanoparticles for improved topical application of drugs for skin diseases. Adv Drug Deliv Rev 59(6):427–443
58. Stein C, Clark JD, Oh U, Vasko MR, Wilcox GL, Overland AC, Vanderah TW, Spencer RH (2009) Peripheral mechanisms of pain and analgesia. Brain Res Rev 60(1):90–113
59. Mine S, Fortunel NO, Pageon H, Asselineau D (2008) Aging alters functionally human dermal papillary fibroblasts but not reticular fibroblasts: a new view of skin morphogenesis and aging. PLoS ONE 3(12):e4066

60. Asselineau D, Ricois S, Pageon H, Zucchi H, Girardeau-Hubert S, Deneuville C, Haydont V, Neiveyans V, Lorthois I (2017) Reconstructed skin to create *In vitro*. In: Flexible models of skin aging: new results and prospects, textbook of aging skin. Springer, Heidelberg, Berlin, pp 1203–1228

61. Pageon H, Zucchi H, Asselineau D (2012) Distinct and complementary roles of papillary and reticular fibroblasts in skin morphogenesis and homeostasis. Eur J Dermatol 22(3):324–332

62. Diekmann J, Alili L, Scholz O, Giesen M, Holtkötter O, Brenneisen P (2016) A three-dimensional skin equivalent reflecting some aspects of *in vivo* aged skin. Exp Dermatol 25(1):56–61

63. Weinmüllner R, Zbiral B, Becirovic A, Stelzer EM, Nagelreiter F, Schosserer M, Lämmermann I, Liendl L, Lang M, Terlecki-Zaniewicz L (2020) Organotypic human skin culture models constructed with senescent fibroblasts show hallmarks of skin aging. NPJ Aging Mech Dis 6(1):1–7

64. Nguyen DG, Pentoney SL Jr (2017) Bioprinted three dimensional human tissues for toxicology and disease modeling. Drug Discov Today Technol 23:37–44

65. Liu X, Michael S, Bharti K, Ferrer M, Song MJ (2020) A biofabricated vascularized skin model of atopic dermatitis for preclinical studies. Biofabrication 12(3):035002

66. Huet F, Severino-Freire M, Chéret J, Gouin O, Praneuf J, Pierre O, Misery L, Le Gall-Ianotto C (2018) Reconstructed human epidermis for in vitro studies on atopic dermatitis: a review. J Dermatol Sci 89(3):213–218

67. Yamanaka K-I, Mizutani H (2011) The role of cytokines/chemokines in the pathogenesis of atopic dermatitis, pathogenesis and management of atopic. Dermatitis 41:80–92

68. Gouin O, Lebonvallet N, L'Herondelle K, Le Gall-Ianotto C, Buhé V, Plée-Gautier E, Carré JL, Lefeuvre L, Misery L (2015) Self-maintenance of neurogenic inflammation contributes to a vicious cycle in skin. Exp Dermatol 24(10):723–726

69. Castex-Rizzi N, Galliano M, Aries M, Hernandez-Pigeon H, Vaissiere C, Delga H, Caruana A, Carrasco C, Lévêque M, Duplan H (2014) *In vitro* approaches to pharmacological screening in the field of atopic dermatitis. Br J Dermatol 170:12–18

70. Pendaries V, Le Lamer M, Cau L, Hansmann B, Malaisse J, Kezic S, Serre G, Simon M (2015) In a three-dimensional reconstructed human epidermis filaggrin-2 is essential for proper cornification. Cell Death Dis 6(2):e1656–e1656

71. Pendaries V, Malaisse J, Pellerin L, Le Lamer M, Nachat R, Kezic S, Schmitt A-M, Paul C, Poumay Y, Serre G (2014) Knockdown of filaggrin in a three-dimensional reconstructed human epidermis impairs keratinocyte differentiation. J Investig Dermatol 134(12):2938–2946

72. Rouaud-Tinguely P, Boudier D, Marchand L, Barruche V, Bordes S, Coppin H, Roth M, Closs B (2015) From the morphological to the transcriptomic characterization of a compromised three-dimensional *in vitro* model mimicking atopic dermatitis. Br J Dermatol 173(4):1006–1014

73. Lebonvallet N, Jeanmaire C, Danoux L, Sibille P, Pauly G, Misery L (2010) The evolution and use of skin explants: potential and limitations for dermatological research. Eur J Dermatol 20(6):671–684

74. De Vuyst É, Giltaire S, Lambert de Rouvroit C, Malaisse J, Mound A, Bourtembourg M, Poumay Y, Nikkels A, Chrétien A, Salmon M (2018) Methyl-beta-cyclodextrin concurs with interleukin (IL)-4, IL-13 and IL-25 to induce alterations reminiscent of atopic dermatitis in reconstructed human epidermis. Exp Dermatol 27(4):435–437

75. do Nascimento Pedrosa T, De Vuyst E, Mound A, De Rouvroit CL, Maria-Engler SS, Poumay Y (2017) Methyl-β-cyclodextrin treatment combined to incubation with interleukin-4 reproduces major features of atopic dermatitis in a 3D-culture model, Archives of dermatological research 309(1):63–69

76. Rioux G, Pouliot-Bérubé C, Simard M, Benhassine M, Soucy J, Guérin SL, Pouliot R (2018) The tissue-engineered human psoriatic skin substitute: a valuable in vitro model to identify genes with altered expression in lesional psoriasis. Int J Mol Sci 19(10):2923

77. Jean J, Lapointe M, Soucy J, Pouliot R (2009) Development of an *in vitro* psoriatic skin model by tissue engineering. J Dermatol Sci 53(1):19–25
78. Tjabringa G, Bergers M, van Rens D, de Boer R, Lamme E, Schalkwijk J (2008) Development and validation of human psoriatic skin equivalents. Am J Pathol 173(3):815–823
79. Barker CL, McHale MT, Gillies AK, Waller J, Pearce DM, Osborne J, Hutchinson PE, Smith GM, Pringle JH (2004) The development and characterization of an in vitro model of psoriasis. J Investig Dermatol 123(5):892–901
80. Slominski A, Zbytek B, Szczesniewski A, Wortsman J (2006) Cultured human dermal fibroblasts do produce cortisol. J Invest Dermatol 126(5):1177
81. Hannen RF, Michael AE, Jaulim A, Bhogal R, Burrin JM, Philpott MP (2011) Steroid synthesis by primary human keratinocytes; implications for skin disease. Biochem Biophys Res Commun 404(1):62–67
82. Korting H, Patzak U, Schaller M, Maibach H (1998) A model of human cutaneous candidosis based on reconstructed human epidermis for the light and electron microscopic study of pathogenesis and treatment. J Infect 36(3):259–267
83. Kühbacher A, Henkel H, Stevens P, Grumaz C, Finkelmeier D, Burger-Kentischer A, Sohn K, Rupp S (2017) Central role for dermal fibroblasts in skin model protection against Candida albicans. J Infect Dis 215(11):1742–1752
84. Schaller M, Preidel H, Januschke E, Korting H (1999) Light and electron microscopic findings in a model of human cutaneous candidosis based on reconstructed human epidermis following the topical application of different econazole formulations. J Drug Target 6(5):361–372
85. Korting H, Klövekorn W, Klövekorn G (1997) Comparative efficacy and tolerability of econazole liposomal gel 1%, branded econazole conventional cream 1% and generic clotrimazole cream 1% in tinea pedis. Clin Drug Investig 14(4):286–293
86. Emmert H, Rademacher F, Gläser R, Harder J (2020) Skin microbiota analysis in human 3D skin models—"Free your mice." Exp Dermatol 29(11):1133–1139
87. Rademacher F, Simanski M, Gläser R, Harder J (2018) Skin microbiota and human 3D skin models. Exp Dermatol 27(5):489–494
88. van Drongelen V, Haisma EM, Out-Luiting JJ, Nibbering P, El Ghalbzouri A (2014) Reduced filaggrin expression is accompanied by increased Staphylococcus aureus colonization of epidermal skin models. Clin Exp Allergy 44(12):1515–1524
89. Lerebour G, Cupferman S, Bellon-Fontaine M (2004) Adhesion of Staphylococcus aureus and Staphylococcus epidermidis to the Episkin® reconstructed epidermis model and to an inert 304 stainless steel substrate. J Appl Microbiol 97(1):7–16
90. Schaller M, Laude J, Bodewaldt H, Hamm G, Korting H (2004) Toxicity and antimicrobial activity of a hydrocolloid dressing containing silver particles in an *ex vivo* model of cutaneous infection. Skin Pharmacol Physiol 17(1):31–36
91. Rebecca VW, Somasundaram R, Herlyn M (2020) Pre-clinical modeling of cutaneous melanoma. Nat Commun 11(1):1–9
92. Hill DS, Robinson ND, Caley MP, Chen M, O'Toole EA, Armstrong JL, Przyborski S, Lovat PE (2015) A novel fully humanized 3D skin equivalent to model early melanoma invasion. Mol Cancer Ther 14(11):2665–2673
93. Marconi A, Quadri M, Saltari A, Pincelli C (2018) Progress in melanoma modelling *in vitro*. Exp Dermatol 27(5):578–586
94. Vörsmann H, Groeber F, Walles H, Busch S, Beissert S, Walczak H, Kulms D (2013) Development of a human three-dimensional organotypic skin-melanoma spheroid model for *in vitro* drug testing. Cell Death Dis 4(7):e719–e719
95. Murekatete B, Shokoohmand A, McGovern J, Mohanty L, Meinert C, Hollier BG, Zippelius A, Upton Z, Kashyap AS (2018) Targeting insulin-like growth factor-i and extracellular matrix interactions in melanoma progression. Sci Rep 8(1):1–12
96. Bourland J, Fradette J, Auger FA (2018) Tissue-engineered 3D melanoma model with blood and lymphatic capillaries for drug development. Sci Rep 8(1):1–13
97. Kaur A, Ecker BL, Douglass SM, Kugel CH, Webster MR, Almeida FV, Somasundaram R, Hayden J, Ban E, Ahmadzadeh H (2019) Remodeling of the collagen matrix in aging skin promotes melanoma metastasis and affects immune cell motility. Cancer Discov 9(1):64–81

98. Li L, Fukunaga-Kalabis M, Herlyn M (2011) The three-dimensional human skin reconstruct model: a tool to study normal skin and melanoma progression, JoVE (J Visualized Exp) 54:e2937

99. Zhang Z, Michniak-Kohn BB (2012) Tissue engineered human skin equivalents. Pharmaceutics 4(1):26–41

100. Abd E, Yousef SA, Pastore MN, Telaprolu K, Mohammed YH, Namjoshi S, Grice JE, Roberts MS (2016) Skin models for the testing of transdermal drugs. Clin Pharmacol: Adv Appl 8:163

101. Pellevoisin C, Bouez C, Cotovio J (2018) Cosmetic industry requirements regarding skin models for cosmetic testing, skin tissue models, Elsevier, pp 3–37

102. Rittié L (2016) Cellular mechanisms of skin repair in humans and other mammals. J Cell Commun Signaling 10(2):103–120

103. De Wever B, Kurdykowski S, Descargues P (2015) Human skin models for research applications in pharmacology and toxicology: introducing NativeSkin®, the "missing link" bridging cell culture and/or reconstructed skin models and human clinical testing. Appl in Vitro Toxicol 1(1):26–32

104. Draelos ZD (2018) The science behind skin care: cleansers. J Cosmet Dermatol 17(1):8–14

105. Zhang Q, Sito L, Mao M, He J, Zhang YS, Zhao XJ (2018) Current advances in skin-on-a-chip models for drug testing. Microphysiol Syst 2

106. Robinson MK, Osborne R, Perkins MA (1999) Strategies for the assessment of acute skin irritation potential. J Pharmacol Toxicol Methods 42(1):1–9

107. Yun YE, Jung YJ, Choi YJ, Choi JS, Cho YW (2018) Artificial skin models for animal-free testing. J Pharm Investig 48(2):215–223

108. Williams AC, Barry BW (2012) Penetration enhancers. Adv Drug Deliv Rev 64:128–137

109. Kansy M, Senner F, Gubernator K (1998) Physicochemical high throughput screening: parallel artificial membrane permeation assay in the description of passive absorption processes. J Med Chem 41(7):1007–1010

110. Ottaviani G, Martel S, Carrupt P-A (2006) Parallel artificial membrane permeability assay: a new membrane for the fast prediction of passive human skin permeability. J Med Chem 49(13):3948–3954

111. Sinkó B, Garrigues TM, Balogh GT, Nagy ZK, Tsinman O, Avdeef A, Takács-Novák K (2012) Skin–PAMPA: a new method for fast prediction of skin penetration. Eur J Pharm Sci 45(5):698–707

112. Tsinman K, Sinko B (2013) A high throughput method to predict skin penetration and screen topical formulations. Cosmet Toiletries 128(3):192–199

113. Flaten G, Palac Z, Engesland A, Filipović-Grčić J, Vanić Ž, Škalko-Basnet N (2015) vitro skin models as a tool in optimization of drug formulation. Eur J Pharm Sci 75:10–24

114. Engesland A, Skar M, Hansen T, Škalko-Basnet N, Flaten GE (2013) New applications of phospholipid vesicle-based permeation assay: permeation model mimicking skin barrier. J Pharm Sci 102(5):1588–1600

115. Engesland A, Škalko-Basnet N, Flaten GE (2015) Phospholipid vesicle-based permeation assay and EpiSkin® in assessment of drug therapies destined for skin administration. J Pharm Sci 104(3):1119–1127

116. Palac Z, Engesland A, Flaten GE, Škalko-Basnet N, Filipović-Grčić J, Vanić Ž (2014) Liposomes for (trans) dermal drug delivery: the skin-PVPA as a novel in vitro stratum corneum model in formulation development. J Liposome Res 24(4):313–322

117. Schäfer-Korting M, Bock U, Gamer A, Haberland A, Haltner-Ukomadu E, Kaca M, Kamp H, Kietzmann M, Korting HC, Krächter H-U (2006) Reconstructed human epidermis for skin absorption testing: results of the German prevalidation study. Altern Lab Anim 34(3):283–294

118. Schäfer-Korting M, Bock U, Diembeck W, Düsing H-J, Gamer A, Haltner-Ukomadu E, Hoffmann C, Kaca M, Kamp H, Kersen S (2008) The use of reconstructed human epidermis for skin absorption testing: results of the validation study. Altern Lab Anim 36(2):161–187

119. Schmook FP, Meingassner JG, Billich A (2001) Comparison of human skin or epidermis models with human and animal skin in in-vitro percutaneous absorption. Int J Pharm 215(1–2):51–56

120. Schreiber S, Mahmoud A, Vuia A, Rübbelke M, Schmidt E, Schaller M, Kandarova H, Haberland A, Schäfer U, Bock U (2005) Reconstructed epidermis versus human and animal skin in skin absorption studies. Toxicol In Vitro 19(6):813–822

121. Gregoire S, Patouillet C, Noe C, Fossa I, Kieffer FB, Ribaud C (2008) Improvement of the experimental setup for skin absorption screening studies with reconstructed skin EPISKIN®. Skin Pharmacol Physiol 21(2):89–97

122. Gabbanini S, Lucchi E, Carli M, Berlini E, Minghetti A, Valgimigli L (2009) In vitro evaluation of the permeation through reconstructed human epidermis of essentials oils from cosmetic formulations. J Pharm Biomed Anal 50(3):370–376

123. Rozman B, Gasperlin M, Tinois-Tessoneaud E, Pirot F, Falson F (2009) Simultaneous absorption of vitamins C and E from topical microemulsions using reconstructed human epidermis as a skin model. Eur J Pharm Biopharm 72(1):69–75

124. Traynor MJ, Wilkinson SC, Williams FM (2007) The influence of water mixtures on the dermal absorption of glycol ethers. Toxicol Appl Pharmacol 218(2):128–134

125. Dong P, Nikolaev V, Kröger M, Zoschke C, Darvin ME, Witzel C, Lademann J, Patzelt A, Schäfer-Korting M, Meinke MC (2020) Barrier-disrupted skin: quantitative analysis of tape and cyanoacrylate stripping efficiency by multiphoton tomography. Int J Pharm 574:118843

126. Krutmann J, Merk HF (2018) Environment and skin. Springer

127. Oliver G, Pemberton M, Rhodes C (1986) An in vitro skin corrosivity test—Modifications and validation. Food Chem Toxicol 24(6–7):507–512

128. Kandárová H, Liebsch M, Spielmann H, Genschow E, Schmidt E, Traue D, Guest R, Whittingham A, Warren N, Gamer AO (2006) Assessment of the human epidermis model SkinEthic RHE for in vitro skin corrosion testing of chemicals according to new OECD TG 431. Toxicol In Vitro 20(5):547–559

129. Bernhofer L, Barkovic S, Appa Y, Martin K (1999) IL-1α and IL-1ra secretion from epidermal equivalents and the prediction of the irritation potential of mild soap and surfactant-based consumer products. Toxicol In Vitro 13(2):231–239

130. Cannon C, Neal P, Southee J, Kubilus J, Klausner M (1994) New epidermal model for dermal irritancy testing. Toxicol In Vitro 8(4):889–891

131. Kubilus J, Cannon C, Neal P, Sennott H, Klausner M (1996) Response of the EpiDerm skin model to topically applied irritants and allergens. In Vitro Toxicol 9(2):157–166

132. Augustin C, Collombel C, Damour O (1998) Use of dermal equivalent and skin equivalent models for in vitro cutaneous irritation testing of cosmetic products: comparison with in vivo human data. J Toxicol: Cutan Ocul Toxicol 17(1):5–17

133. Roguet R, Cohen C, Robles C, Courtellemont P, Tolle M, Guillot J, Duteil XP (1998) An interlaboratory study of the reproducibility and relevance of Episkin, a reconstructed human epidermis, in the assessment of cosmetics irritancy. Toxicol In Vitro 12(3):295–304

134. Alepee N, Tornier C, Robert C, Amsellem C, Roux M, Doucet O, Pachot J, Meloni M, de A Brugerolle de Fraissinette (2010) A catch-up validation study on reconstructed human epidermis (SkinEthic RHE) for full replacement of the Draize skin irritation test. Toxicol In vitro 24(1):257–266

135. Pfuhler S, Fellows M, van Benthem J, Corvi R, Curren R, Dearfield K, Fowler P, Frötschl R, Elhajouji A, Le Hégarat L (2011) In vitro genotoxicity test approaches with better predictivity: summary of an IWGT workshop. Mutat Res/Genet Toxicol Environ Mutagenesis 723(2):101–107

136. Kirsch-Volders M, Decordier I, Elhajouji A, Plas G, Aardema MJ, Fenech M (2011) In vitro genotoxicity testing using the micronucleus assay in cell lines, human lymphocytes and 3D human skin models. Mutagenesis 26(1):177–184

137. Ghosh A, Roth K (2014) Detection of apoptosis and autophagy

138. Flamand N, Marrot L, Belaidi J-P, Bourouf L, Dourille E, Feltes M, Meunier J-R (2006) Development of genotoxicity test procedures with Episkin®, a reconstructed human skin model: towards new tools for in vitro risk assessment of dermally applied compounds? Mutat Res/Genet Toxicol Environ Mutagenesis 606(1–2):39–51

Chapter 7
Conclusions

The translation of tissue engineering concepts into the development of biomimetic skin models is a growing trend that began as a treatment strategy to save and improve patients' quality of life. As summarized in this book, various ("in-house" and commercial) in vitro skin models have attempted to fully replicate the native tissue while meeting the needs of various industries, from clinics to cosmetics and pharmaceutics. It is, however, also grown for its own merit and has already gained a crucial position in regenerative medicine. Huge investments have boosted its continuous development. These financial investments of prominent cosmetic companies like L'oréal improved the existing skin models. The synergistic advances in cell cultivation, materials science, and bioengineering not only enable the establishment of more sophisticated healthy and diseased skin models with improved structural and functional accuracy but have also greatly influenced all other branches of tissue engineering, which built on findings based on the increasing complexity of artificial skin products. The introduction of new strategies of micro- and nano-fabrication, biomaterial synthesis, functionalization techniques, and the use of patient-specific cells have provided a framework to produce remarkably potent functional skin substitutes that could overcome the challenges of currently available skin grafts. In addition, they have become an integral part of in vitro testing and evaluation of skin care products and therapeutic applications. In clinical applications, additional artificial skin features are sought, for example, minimization (and ideally complete elimination) of scarring, as well as the need for broad efficacy in a variety of patient populations and wound types. However, the clinical value of novel models is not limited to anatomical reconstruction but is increasingly focused on models that can elucidate the mechanisms behind various skin pathologies. The potential for engineering complex skin equivalents that actually mimic the microenvironment in vivo has expanded greatly in recent decades. Many new strategies and viable technologies are already reaching the product market. They contain certain synthetic components along with natural components that can provide adequate ECM components and can also be used to form dermal equivalents. However, in such culture systems, the communications between different cell types of dermal and epidermal layers are

T. Zidarič et al., *Function-Oriented Bioengineered Skin Equivalents*,
Biobased Polymers, https://doi.org/10.1007/978-3-031-21298-7_7

difficult to reproduce, limiting the functionality of these artificially produced skin products. Skin models on the market today are largely representative of the epidermis, which presents a need for further development. For example, improved barrier function in the *stratum corneum*, better representation of the epidermal-dermal junction mirroring the viscoelastic properties of the skin, and the inclusion of accurate immune response are possible future directions for the development of improved skin models. Moreover, in the fabrication of 3D skin, we still face many of the same problems that preoccupied scientists in the mid-twentieth century, such as physiological oxygen and nutrient delivery with a perfused vasculature, complex structures such as glands and hair follicles. Biofabrication techniques (3D bioprinting and electrospinning) not only allow for the straightforward generation of such constructs but also hold promise for the fabrication of appendages and macrostructures in skin tissue containing these specialized cell types for the formation of glands and hair follicles. All of this will enhance the ability to produce physiological skin for personalized medicine and other preclinical and commercial applications (cosmetic and skin care product testing, drug screening). The production of the necessary physiologically relevant skin components, from the ECM to the microbiome, is feasible with current bioengineering technologies. However, additional advances in existing technologies and the development of entirely new technologies will enable cost-effective and reproducible generation of physiological skin in vitro. Once developed, skin equivalents will greatly improve patient quality of life, reduce reliance on current animal models and provide an accurate and relatively definitive screening tool for product testing.

Acknowledgement The authors acknowledge the financial support from the Slovenian Research Agency (grant numbers: P3-0036, I0-0029, J3-1762, J7-4492 and L7-4944).

Index

Printed in the United States
by Baker & Taylor Publisher Services